# THERMODYNAMICS OF STRUCTURE

# Structure and Properties of Materials

---

VOLUME I     STRUCTURE

William G. Moffatt
George W. Pearsall
John Wulff


VOLUME II     THERMODYNAMICS OF STRUCTURE

Jere H. Brophy
Robert M. Rose
John Wulff


VOLUME III     MECHANICAL BEHAVIOR

H. Wayne Hayden
William G. Moffatt
John Wulff


VOLUME IV     ELECTRONIC PROPERTIES

Leander F. Pease
Robert M. Rose
John Wulff

# THERMODYNAMICS OF STRUCTURE

*Jere H. Brophy*

*Robert M. Rose*

*John Wulff*

JOHN WILEY & SONS, INC. *New York · London · Sydney*

# Preface

The four brief volumes in this series were designed as a text for a two-semester introductory course in materials for engineering and science majors at the sophomore-junior level. Some curricula provide only one semester for a materials course We have found that under such circumstances it is convenient to use Volumes I, II, and parts of III for aeronautical, chemical, civil, marine, and mechanical engineers. Similarly, parts of Volumes I, II, and IV form the basis for a single-semester course for electrical engineering and science majors.

The four volumes grew from sets of notes written for service courses during the last decade. In rewriting these for publication, we have endeavored to emphasize those principles which relate properties and behavior of different classes of materials to their structure and environment. In order to develop a coherent and logical presentation in as brief a context as possible, we have used problem sets at the end of each chapter to extend and illustrate particular aspects of the subject. Real materials encountered in engineering situations have been chosen as examples wherever possible. Problem and laboratory sections to supplement lectures have aided our students considerably in applying the principles delineated in the text to a variety of materials and environments.

Many of the tables and illustrations used in the present text have been borrowed from individual specialists and their publications. Our thanks are due both to the individuals responsible for the data and to the publishers. Their names are listed with the illustrations in the text. Further thanks are due to numerous colleagues who took part in teaching the same courses with us during the past ten years. Many parts of the text have been improved as a result of their constructive criticism.

Finally, we wish to acknowledge our indebtedness to the Ford Foundation and to Dr. Gordon S. Brown, Dean of Engineering at M.I.T., who early supported our efforts to provide lecture

vi     Preface

demonstrations, laboratory experiments, and notebook editions
of the present text for the use of our students.

*July 1964*

JERE H. BROPHY, International Nickel Company, Inc.
ROBERT M. ROSE, Massachusetts Institute of Technology
JOHN WULFF, Massachusetts Institute of Technology

# *Contents*

# THERMODYNAMICS OF STRUCTURE

CHAPTER ONE

# Thermodynamics

SUMMARY

The laws of thermodynamics are generalizations of common experience. Simple measurements may be made of pressure, volume, temperature, chemical composition, and other appropriate quantities; such data specify the *state* of the object or region of interest (called the *system*) and all its properties. If a system is left undisturbed, it will eventually come to equilibrium, and all its properties will no longer vary with time. The First Law of Thermodynamics is the statement of the conservation of energy, which leads to the conclusion that the internal energy of any system is a function only of the thermodynamic state. Although heat and work are equivalent forms of energy, they are not completely interchangeable. According to the Second Law, heat may never be entirely converted to work, and heat never changes spontaneously to work. Such behavior is described by the maximization of a new state function, the *entropy* which increases as a system approaches equilibrium and is maximized at equilibrium. Such quantities are relatively easy to measure and permit the prediction of the equilibrium state of any thermodynamic system. The spontaneous occurrence of a natural event can then be described in terms of a decrease in the appropriate free energy function.

## 1.1 INTRODUCTION

The basis of thermodynamics is common, everyday experience. The laws of thermodynamics are, it is true, exact and mathematical in nature; however, each law follows from simple and direct observation of natural events. By way of example, consider what we must know in order to specify the properties of such complicated systems as a solid object or a body of gas. If the position and velocity of every atom therein were precisely known, and all interatomic and other forces specified, and all of the properties of

1

individual atoms known, it might be possible, by tremendous labor, to calculate or predict all of the physical properties of whatever portion of the universe happened to be of interest. This portion of the universe is, incidentally, generally referred to as a *system*. Thermodynamics is not concerned with such atomic or microscopic detail but deals instead with macroscopic, directly measurable quantities. Pressure, volume, temperature, and chemical composition are typical of the macroscopic quantities. If necessary, fields, whether gravitational or electromagnetic, may also be measured. In solids, the mechanics of elasticity and plasticity are important.

It is always possible, if the set of measurements is complete enough, to determine all other physical properties of the system. Such a set of numbers is said to fix the thermodynamic state of the system. With each measurement, or set of measurements, an equilibrium is associated. If, for instance, no motion occurs, owing to complete balancing of forces, mechanical equilibrium has been attained. If the chemical composition does not change with time anywhere, chemical equilibrium has occurred. Thermal equilibrium is associated with the absence of heat flow because temperature is everywhere equalized. When all of the variables which specify the thermodynamic state are unchanging, the system is in thermodynamic equilibrium.

If our measurements of the state of any system are to be free of embarrassing uncertainties, the state involved should be an equilibrium state. Observations made "in transit" are likely to depend on where we have been, rather than where we are. This dependence upon past history is common in nonequilibrium situations, which are by far the most frequently encountered. After all, equilibrium is by definition the last state arrived at by an unperturbed system; it is a universal rarity. It is important only as a guide to behavior, that is, as an indication of the direction in which natural events move. This direction is really the fundamental problem of thermodynamics. Given a certain set of initial conditions and the knowledge that the system will spontaneously move towards equilibrium, one must predict the equilibrium state. The experimental and analytical determination of quantities that define the state and changes in states provide the principal tools of thermodynamics.

## 1.2  ENERGY CONSERVATION, STATE FUNCTIONS, AND THE FIRST LAW

We will assume a certain familiarity with the nature of energy and the fact that it may exist in the equivalent forms of heat and work. The most familiar principle involving energy is that energy must be conserved. Such a statement, however, leads immediately to the following conclusion: the internal energy content of any system is a single-valued function only of the thermodynamic state.

If this conclusion were not so, contradiction of the energy-conservation principle would be a simple matter. A system is chosen which is relatively isolated from the rest of the universe; a change of state is effected and the change in energy content noted. The system is then returned to its initial state by a completely different method. The net change of the internal energy on completion of the "cycle" must be zero; otherwise energy would have appeared or disappeared, in contradiction to a principle previously agreed upon. The methods by which the changes in state were accomplished are completely arbitrary and, therefore, the energy content change which accompanies any change of state must not depend on the path, but only on the initial and final states, whether they are equilibrium states or nonequilibrium states. Similar reasoning applies to any conserved quantity, which is then referred to as a *state function*.

For any system which undergoes a change in state, we may then write

$$\Delta E = q - w \qquad (1.1)$$

where $E$ is the internal energy, $q$ the total heat input to the system, and $w$ the total work output during the change. The First Law states that the change in internal energy corresponding to a change of state between stipulated initial and final states does not depend on the method used to achieve the transformation. The quantities $q$ and $w$ are not properties of the system and will depend on methods; $\Delta E$, however, is a function only of the states, and is therefore independent of the path between them.

A particularly convenient change of state to observe is the rise in temperature due to input of heat. Such observations made under common laboratory conditions lead to the definition

$$C_p \equiv \left(\frac{\partial q}{\partial T}\right)_P \tag{1.2}$$

where $C_p$ is called the heat capacity at constant pressure. No linear relationship is implied between $q$ and $T$ by equation 1.2; $C_p$ is merely the instantaneous slope of a more complicated curve. It is useful to connect such experimental quantities with the state functions. If only the atmosphere does work on our system, the differential of $E$ is then

$$dE = dq - P\,dV \tag{1.3}$$

If the volume is held constant, then

$$\left[\frac{\partial E}{\partial T}\right]_V = \left[\frac{dq}{dT}\right]_V = C_V \tag{1.4}$$

where $C_V$ is called the heat capacity at constant volume. It is usually difficult, however, to hold the volume constant. Consequently, it is convenient to define a new state function, the *enthalpy*:

$$H \equiv E + PV \tag{1.5}$$

Because the thermodynamic state includes values of $P$ and $V$, and $E$ is a state function, $H$ is also a single-valued function of state. The differential of $H$ at constant pressure leads immediately to

$$\left[\frac{\partial H}{\partial T}\right]_P = \left[\frac{dq}{dT}\right]_P = C_p \tag{1.6}$$

The enthalpy, which is also conserved, is therefore a more useful state function in some ways than the energy, as it is more directly related to our measurements.

## 1.3   THE SECOND LAW

The previous section refers briefly to the equivalence of heat and work as forms of energy. Experience shows, however, that such a statement is not completely and unconditionally acceptable. If heat and work were always completely interchangeable, truly remarkable events might occur. Stones, cooling themselves by converting their heat contents to potential energy (i.e., work), could roll uphill unaided; one could never trust hot water to "stay put," as it might at any time convert its heat to kinetic or potential energy, with uncomfortable results.  In general then, nature does not allow heat to be spontaneously converted to work without the accompaniment of other changes.  This last statement is actually Kelvin's formulation of the Second Law.  Steam, gasoline, and other engines all convert heat to work; what "other changes" occur during such conversion?

For the sake of simplicity, friction should be eliminated from consideration.  As a limiting case, the idea of a *reversible process*, or a reversible change, is devised.  The condition of reversibility is exactly what the term implies.  A reversible process may be exactly retraced, or reversed, to its beginning.  The process may be carried out backwards and everything restored to its original state. *Dissipation*, or friction, would then be out of the question, as the energy dissipated is irrecoverable.  A pendulum, for instance, can retrace its path exactly only if friction is absent. Reversible processes operate with maximum efficiency.  If a number of different reversible methods accomplish the same change of state, they must all be equally efficient.  Even in the absence of obvious sources of friction, a process can be reversible only if the system is, at all times, very close to equilibrium.  The situation must be one of *virtual equilibrium;* the process is a series of very minute (ideally infinitesimal) perturbations from equilibrium. Two examples, comparisons of reversible and irreversible means of accomplishing the same processes, are seen in Figures 1.1 and 1.2.

Temporarily then, friction may be dropped from the discussion, and the reversible conversion of heat to work analyzed.  It is simplest to utilize a hypothetical engine, which absorbs heat, does work, and is then restored to its initial state, all in a reversible manner.  The changes occurring over the entire cycle may be regarded as those attendant on the conversion of heat to work in the most efficient manner possible.  Detailed analyses of reversible heat engines of this sort will be found in many of the references at the end of this chapter.  We will use only the results.  Of primary interest is the theorem of Clausius:

$$\oint \frac{dq}{T} \leq 0 \qquad (1.7)$$

**Reversible**

Very slowly outward

$P_{\text{inside}} - \delta P$

$P_{\text{inside}}$

$dw_{\text{rev.}} = P_{\text{rev.}} \, dV$

$P_{\text{rev.}} = P_{\text{inside}}$

$w_{\text{rev.}} = \int_{P_{\text{initial}}}^{P_{\text{final}}} P_{\text{inside}} \, dV$

$PV = NRT$

$w_{\text{rev.}} = NRT \ln\left(\dfrac{P_{\text{initial}}}{P_{\text{final}}}\right)$

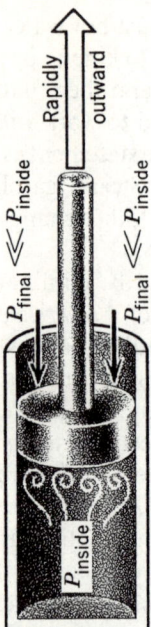

**Irreversible**

Rapidly outward

$P_{\text{final}} \ll P_{\text{inside}}$

$P_{\text{inside}}$

$dw_{\text{irrev.}} = P_{\text{irrev.}} \, dV$

$P_{\text{irrev.}} = P_{\text{inside}} - P_{\text{turbulence}}$

$w_{\text{irrev.}} = \int_{P_{\text{initial}}}^{P_{\text{final}}} P_{\text{irrev.}} \, dV$

$P_{\text{irrev.}} < P_{\text{rev.}}$

$w_{\text{irrev.}} < w_{\text{rev.}}$

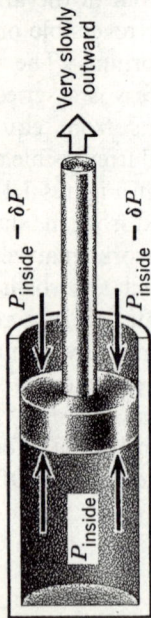

$P^{\text{rev.}}_{\text{irrev.}}$ = pressure acting <u>on cylinder</u>

Figure 1.1 Reversible and irreversible expansion.

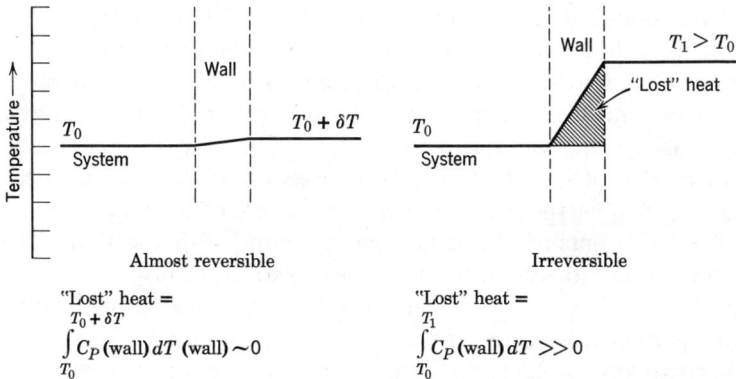

Figure 1.2 Reversible and irreversible heat flow.

The integral is taken around one complete cycle. The inequality applies to irreversible cycles, and the equality to reversible cycles. $T$ is an "absolute" temperature. We could not, according to Clausius' Theorem, just extract heat and convert it to work. During the cycle it would be necessary to reject heat as well in order to satisfy equation 1.7. The rejected heat, of course, represents that portion of the heat intake which cannot be converted to work. Also, rejection of heat must take place at lower temperatures than the heat intake, if work is to be done. This is easily seen. To satisfy the First Law (equation 1.1), the heat input minus the heat output must be equal to the work done. If the work is positive, the heat output is always less than the original heat input. Consequently, the range of temperatures over which heat rejection occurs is, in general, lower than the temperature range where heat is absorbed in order for equation 1.7 to be satisfied. If heat is absorbed at a single temperature, and no lower temperature is available, equation 1.7 requires that the same heat be rejected to that temperature, and no work done. Thus, one cannot convert to work the large amounts of heat energy available in the oceans, for instance, as no large, low temperature "reservoirs" exist for the dumping of the amounts of heat required by equation 1.7. The inequality in the above example corresponds to the lower efficiency of irreversible methods; more of the heat taken in must be rejected, and consequently less converted to work.

We should not lose sight of our intention. Equation 1.7 is not intended to be an "explanation" of the natural phenomena we have discussed, but rather a description. The scope of this description may now be expanded greatly by the introduction of a new thermodynamic variable, the *entropy*. Thus far, the purposeful conversion of heat to work, by various cyclic devices, has been analyzed; now, the second law must be extended to natural, "spontaneous" events mentioned at the beginning of this section. The simplest way to accomplish the above, is to use Clausius' Theorem to define a new state function. For a reversible cycle, the integral in equation 1.7 is zero. Two thermodynamic states connected by *any* two reversible paths constitute such a cycle, regardless of path. Therefore, the integral

$$\int_A^B \frac{dq_{\text{rev.}}}{T}$$

connecting the states $B$ and $A$ is a function only of the initial and final states, and *not* the paths connecting them. Consequently, the differential of a new state function

$$dS \equiv \frac{dq_{\text{rev.}}}{T} \tag{1.8}$$

is being integrated above. $S$ is the symbol for the entropy. Also, from equation 1.7, the relation

$$dq_{\text{rev.}} > dq_{\text{irrev.}} \tag{1.9}$$

must apply to reversible and irreversible methods for the same change of state. Consider an arbitrarily small change of state where the heat $dq$ is transferred. By equation 1.9

$$dq \leq dq_{\text{rev.}} \tag{1.9a}$$

The change in entropy of the system, regardless of reversibility, is

$$dS_{\text{sys.}} = \frac{dq_{\text{rev.}}}{T}$$

as entropy is determined by the initial and final states only. Because $dq$ is so small, the surroundings may be returned to their initial state by restoration of $dq$; all other changes are presumed to be second order.  The entropy change of the surroundings may be obtained by restoring $dq$ reversibly:

$$dS_{\text{surr.}} = \frac{-dq}{T}$$

The *net* entropy change during any such differential state change is then

$$dS_{\text{net}} = dS_{\text{sys.}} + dS_{\text{surr.}} = \frac{dq_{\text{rev.}} - dq}{T} \geq 0 \qquad (1.10)$$

Once again, the equality refers to reversibility.  A spontaneously occurring process is by nature irreversible, otherwise it would not have occurred.  Reversibility is attainable only when equilibrium is so near that no observable changes will occur.  According to equation 1.10 then, *any spontaneous (natural) event implies an increase in total entropy.*  The equality in equation 1.10 simply means that *entropy* is *maximized at equilibrium.*

These statements, or the equation, are the most general form of the Second Law.  *It is our experience,* then, that no event occurs spontaneously which decreases entropy, including those of the hot water and the rolling stone, which were discussed at the beginning of this section.

## 1.4  ENTROPY, TEMPERATURE, AND THE THIRD LAW

There is an important omission in the preceding sections; the temperature is undefined.  Pressure, volume, work, heat, and the other variables have definitions which can be carried over from classical mechanics.  To say, "Temperature is what is measured with a thermometer," is unsatisfactory as no ordinary thermometer will measure the absolute temperature, which is the only useful thermodynamic scale.  The only thermodynamically consistent temperature scale is obtained by using the Second Law.  To do this, we consider a reversible, cyclic engine, as in Section 1.3.  This engine, however, has the special ability to absorb heat isothermally at the upper temperature $T_2$ and to reject it isothermally at the lower temperature $T_1$.  The Clausius Theorem, (equation 1.7) applied to such a cycle gives

$$\frac{Q_2}{T_2} + \frac{Q_1}{T_1} = 0 \qquad (1.11)$$

where $Q_2$ is the heat absorbed at temperature $T_2$ and so on. The temperature scale is now established by the ratio of heat input to outflow:

$$\frac{Q_2}{Q_1} = -\frac{T_2}{T_1} \qquad (1.11a)$$

This relation is independent of the manner of construction of the engine. We may now arbitrarily assign a number to some easily recognizable temperature (such as the freezing point of pure water at one atmosphere pressure), and all other such temperatures follow from equation 1.11a.

Having used the absolute temperature to define the Second Law (entropy), we now find the Second Law is the only consistent way to define temperature. There is only one way out; entropy must be accepted as a fundamental variable, as basic in nature as temperature.

In mechanics, for instance, we have a consistent system: length, time, mass, force, momentum, energy. Only one of the last four need be chosen, and the rest can be defined in terms of it, and length and time. Such a choice is arbitrary and is usually made according to the way in which one wishes to see the laws of mechanics formulated. So it is with entropy and temperature. We may choose either to define the other, but neither is to be regarded as "more fundamental."

One more requirement on the entropy must be made. Integration of equation 1.8 will give the entropy, up to some arbitrary constant of integration. In 1906, Nernst, following certain suggestions of Planck, found that at absolute zero the entropies of all systems must be equal. This universal constant has been set equal to zero:

$$[S]_{T=0} = 0 \qquad (1.12)$$

The above relation, sometimes called the Third Law of Thermodynamics, allows the calculation of *absolute entropy* by setting the lower limit of integration at the absolute zero of temperature. The integral itself is defined by combining equations 1.2 and 1.8:

$$dS = \frac{dq_{\text{rev.}}}{T} = \frac{C_p \, dT}{T} \qquad (1.13)$$

and therefore

$$S(T) = \int_0^T \frac{C_p}{T'} dT' \qquad (1.13a)$$

## 1.5   FREE ENERGIES: THE THERMODYNAMIC POTENTIALS

We have now almost done what we set forth to do in the introductory section of this chapter. A new variable, the entropy, has been defined. This variable increases as a system moves toward equilibrium, and it is maximized at equilibrium. Consequently, if we can calculate the entropy of any system, we can also predict what changes will occur and where equilibrium will be reached. As a final step, a criterion for equilibrium will be derived which is in a more familiar form and involves quantities which are more directly measurable. Sometimes, when we seek to compute the *net* change in entropy accompanying an event of interest, we find the coupling of the system to the surroundings so complicated that such computations become laborious, if not impossible. The equilibrium criteria which follow circumvent this problem.

In mechanics, the significance of a minimum in the potential energy function is well understood. Such minima represent equilibrium positions. Similarly, a good "thermodynamic potential" will be minimized at thermodynamic equilibrium. The problem is one of choice. The internal energy is not suitable. We have shown, in the previous sections, that all of the internal energy may not be utilized to do work. Work is really the basis of a well-defined potential energy. We do have, however, an accurate way to calculate how much of the internal energy may not be utilized thus: the entropy. According to the Second Law, if the heat $Q_0 = T_0 \Delta S$ is taken on at temperature $T_0$, with the intention of converting it to work, with all processes carried out in the most efficient (i.e., reversible) manner, we shall still have to reckon with the necessity of dropping off the heat $Q_1 = -T_1 \Delta S$ at some lower temperature $T_1$. Otherwise, the system itself cannot be returned to its original state. As the heat $Q_1$ represents the portion of $Q_0$

which cannot be converted to work, and we can obtain $Q_1$ for any temperature provided the entropy change is known, the entropy is commonly associated with *unavailable* energy. Consequently, as a first try, we consider the function

$$A \equiv E - TS \tag{1.14}$$

$A$ is referred to as the *Helmholtz Free Energy*. If equation 1.14 is differentiated, and the First Law, $dE = dq - dw$, and the definition of entropy (equation 1.8) are both used, the relation

$$dA = dE - T\,dS - S\,dT$$
$$= dq - dq_{\text{rev.}} - dw - S\,dT \tag{1.14a}$$

results. Now, if we stipulate that the temperature be constant, and no work be done on or by the system,

$$[dA]_{w,\,T} = dq - dq_{\text{rev.}}$$

From the Second Law, equation 1.9a states that

$$dq \leq dq_{\text{rev.}}$$

Consequently,

$$[dA]_{w,\,T} \leq 0 \tag{1.15}$$

where the equality applies to equilibrium, and the inequality to spontaneous changes. In an isothermal system on which no work is done, the Helmholtz Free Energy decreases for all natural changes and is minimized at equilibrium. Maintaining constant temperature is a reasonable condition; no work, however, is another story. How, for instance, do we prevent expansion or contraction, which, together with atmospheric pressure, do the work

$$w = \int P\,dV$$

Similar problems exist for changes in magnetic and electric polarization and for many other properties. In Section 1.2, the problem of volume changes was dealt with successfully for the case of heat capacities, by considering enthalpy ($H$) rather than energy. We shall do the same here. The Gibbs Free Energy is defined:

$$F \equiv H - TS = E + PV - TS \tag{1.16}$$

where the last equality comes from equation 1.5, the definition of the enthalpy.  We now subject the Gibbs Free Energy to the same routine that was performed on the Helmholtz Free Energy:

$$dF = dE + P\,dV + V\,dP - T\,dS - S\,dT$$
$$= dq - dw + P\,dV + V\,dP - dq_{\text{rev.}} - S\,dT \quad (1.17)$$

If only the atmosphere does work on the system, $dw = P\,dV$, and

$$dF = dq - dq_{\text{rev.}} + V\,dP - S\,dT \quad (1.18)$$

Now we need only hold pressure and temperature constant, and

$$[dF]_{P,T} = dq - dq_{\text{rev.}} \quad (1.19)$$

and therefore

$$[dF]_{P,T} \leq 0 \quad (1.20)$$

The Gibbs Free Energy is minimized at equilibrium if temperature and pressure are held constant.  Such conditions are relatively easy to attain.  If magnetic, electric, or other fields are present, and cannot be neglected when the work is considered, the situation becomes more complicated.  For our purposes, such necessities will not occur.  The goal is now attained.  From statements (i.e., thermodynamic laws) of universal experience, a state function is derived which decreases for all spontaneous changes and is minimized at thermodynamic equilibrium.  Furthermore, this state function, the Gibbs Free Energy, is minimized under simple conditions and is easy to measure.  Chapter 2 deals with the use of the Gibbs Free Energy.

## DEFINITIONS

*Enthalpy (H).*  The sum of the internal energy and the product of pressure and volume: $H \equiv E + PV$.

*Entropy (S).*  A state function defined thermodynamically by the Clausius Theorem (equation 1.7) and commonly associated with the unavailable energy of a system; maximized at equilibrium.

*Equilibrium.*  The state of a system which prevails when all properties and thermodynamic variables remain constant for indefinite periods of time.

*Gibbs Free Energy (F).*  The enthalpy less the "unavailable energy" $TS$: $F \equiv H - TS$.  Free energies are minimized at equilibrium depending on the constraints on the system.

*Heat Capacity (C).*   The response of a system to heat transfer: $C \equiv \Delta q / \Delta T$.

*Helmholtz Free Energy (A).*   The internal energy of a system minus the "unavailable energy," $TS$: $A \equiv E - TS$.

*Internal Energy (E).*   The total energy within the boundaries of a system; changes in the internal energy are defined by the First Law (equation 1.1).

*Macroscopic.*   Of such size and duration that individual atomic motions are not observable.   Averages or summations, in time and space, of the motions of large numbers of atoms are measured instead.

*Reversible Process:* a dissipation-free change which may be exactly retraced to its initial position.   At all times during the process, the system(s) involved are infinitesimally close to equilibrium (i.e., in *virtual equilibrium*).

*State.*   The specification of the complete physical constitution and properties of a system, associated with a particular set of values for the thermodynamic variables.

*State Function.*   Any quantity which is conserved and consequently must be a single-valued function of the thermodynamic state.   Changes in such a function depend on the initial and final states only, and not on the path between them.

*System.*   The object or region of interest.

## BIBLIOGRAPHY

Daniels, F., and R. A. Alberty, *Physical Chemistry,* Wiley, New York, 1961, second edition.   The first five chapters cover the three laws of thermodynamics from a thermochemical point of view.

Darken, L. S., and R. W. Gurry, *Physical Chemistry of Metals,* McGraw-Hill, New York, 1953.   Chapters 6 to 8 cover the three laws of thermodynamics with experimental examples and illustrations, especially for thermal properties of solids.

Glasstone, S., *Thermodynamics for Chemists,* Van Nostrand, New York, 1947.   Chapters 1 to 5 and 7 cover the same material as Daniels and Alberty from essentially the same point of view.

Guggenheim, E. A., *Thermodynamics,* North-Holland, Amsterdam, 1959.   Pages 1 to 20 cover the First and Second Thermodynamic Laws in the introductory section of this advanced, rigorous, and comprehensive treatment.

Huang, K., *Statistical Mechanics,* Wiley, New York, 1963.   Chapter 1 is a brief, lucid review of thermodynamics for advanced undergraduates or graduate students in physics.

## Problems

1.1    The equation of state for one mole of a "perfect gas" at equilibrium is

$$PV = RT$$

Suppose the gas were expanded reversibly at constant temperature, so that $P$ decreased from $P_1$ to $P_2$.   Use equation 1.3 to find the work done.

1.2   Gay-Lussac (1807) and Joule (1844) found that the energy content of a "perfect gas" depends only on temperature and not on pressure or volume.   Find the total heat transfer $q$ which occurred during the reversible expansion of Problem 1.1.

1.3   Using Gay-Lussac's discovery (Problem 1.2), show that

$$dq = \frac{-RT}{P} dP$$

for reversible expansion of a perfect gas.   Also, show that

$$ds = -\frac{R}{P} dP$$

1.4   Use the result of Problem 1.3 to find the change in entropy of a perfect gas due to reversible expansion from pressure $P_1$ to pressure $P_2$. Suppose the expansion was carried out so irreversibly, as in Figure 1.1, that no work is done (*free expansion*).   What is the change in entropy of the gas now?

1.5   Consider the *total* changes in entropy which occur during the expansions of Problem 1.4.   Show that the reversible expansion does not change the total entropy; find the net entropy change for the free expansion.   Show that the entropy change for free expansion is equal to the work which could have been done if the expansion were reversible, divided by the temperature.   Where has the work gone?   Is it still usable?

1.6   From equation 1.18, prove that

$$\frac{\partial F}{\partial T} = -S$$

1.7   Using the result of Problem 1.6 and equation 1.16, show that

$$\frac{\partial \left( \frac{F}{T} \right)}{\partial T} = -\frac{H}{T^2}$$

1.8   With the solutions of Problems 1.6 and 1.7 and various equations from Chapter 1, suggest at least three alternative ways to derive the free energy of a system from heat-capacity ($C_p$) data.   Is any additional information necessary?

1.9   Use the reversible cyclic engine of Section 1.4 to show that it is impossible to attain the absolute zero.   In particular, let the engine act as a refrigerator, *absorbing* heat at the lower temperature.   Express the work done as a function of the heat absorbed. What happens as the lower temperature approaches zero?

1.10    Deduce from the Second Law why the makers of turbines, rockets, nuclear reactors, and other "heat engines" continually strive for higher operating temperatures.

1.11    From equation 1.18, find the free energy change (per mole) of a perfect gas (see Problem 1.1) due to reversible, isothermal expansion from pressure $P_1$ to pressure $P_2$. Compare with the result of Problem 1.1.

1.12    Show from the Second Law, that objects do not levitate themselves by converting thermal energy to potential energy.

1.13    Show that heat flow from higher to lower temperatures increases entropy. Show that heat does not flow from lower to higher temperatures.

1.14    The specific heat capacity of solid copper is

$$C_p = 5.41 + 1.50 \times 10^{-3}T$$

cal/mole/°K at and above room temperature. Obtain the enthalpy as a function of temperature, using equation 1.6. Use room temperature enthalpy $H(298.16)$, as a reference level, that is, obtain $H(T) - H(298.16)$.

1.15    Using the heat capacity data of the previous question together with equation 1.13, obtain the entropy referred to room temperature, $S(T) - S(298.16)$. Combine this with the result of Problem 1.14 to obtain the free energy, $F(T) - F(298.16)$.

1.16    The heat capacity of liquid copper is 7.50 cal/mole/°K above room temperature. Calculate the enthalpy and entropy referred to room temperature.

1.17    At 298°K the enthalpy of liquid copper is 2200 cal more than that of the solid. The entropy of liquid copper is 0.72 cal/degree/mole more than that of the solid, also at 298°K. Heat capacity data for the solid and liquid, and for the enthalpy and entropy data, are available in Problems 1.16 and 1.15. From this information predict the melting point of pure copper. Do not attempt to solve the equation for the melting point explicitly; solve by trial substitution. Compare the result with the measured melting point, 1357°K.

CHAPTER TWO

# Equilibrium in Multicomponent Systems

SUMMARY

Of the various graphic methods used to indicate the ranges of pressure, temperature, and composition where specific phases and mixtures of phases are stable, the so-called *phase diagram*, which considers temperature and composition variation at one atmosphere pressure, is the most popular and successful. Although phase diagrams are simply records of observations of the phases present, they may also be developed from crude, basic assumptions by applying the free energy concept to solid solutions and mixtures of solid solutions. By considering the influence of atomic, chemical, and crystal properties on the free energy, it is possible to predict approximately the extent of miscibility of two elements. In two-phase regions of binary systems, where the total composition is known, we may use mass conservation to calculate the relative amounts of the two phases present. The use of free energy in analyzing the chemical equilibrium of a multicomponent system leads to the *Phase Rule,* which relates the number of independent thermodynamic variables to the number of components and phases present.

## 2.1 INTRODUCTION

It is possible, without being concerned with atomic detail, to examine the internal features of any system. The liquid and vapor of the same substance (e.g., water and water vapor) remain in equilibrium with each other over a large range of temperatures and pressures. A system like this—one in which the two phases of a given substance (here, the liquid and the vapor) are easily discernible and mechanically separable—is called a *two-phase system.* This terminology may be extended to other systems: liquid-solid

17

(two-phase), liquid-solid-vapor (three-phase), etc. A system containing two immiscible liquids clearly has two liquid phases. Because gases are always completely miscible, only one gaseous phase exists in any system at equilibrium. For instance, a water–ethyl alcohol mixture is only one-phase; a water-oil mixture is two-phase (even if the oil is finely divided, the two-phase structure is readily visible under the microscope); and, of course, any number of gases mix readily and are therefore one-phase. Two solid elements or compounds may mix together readily and thus form a one-phase *solid solution.* On the other hand, differences in crystal structure, valence, or atomic size may prevent the complete interdissolution of the two components, and two solid solutions (i.e., two solid phases) will be formed. Since the two-phase structure is, of course, on the scale of the grain size of the solid, it is readily visible in the optical microscope. In addition, any difference in crystal structure will show up readily in the x-ray diffraction patterns the material produces. The microhardness tester will measure any difference in hardness of the two phases, and once again the two phases are, in principle, mechanically separable.

Any one phase is homogeneous *at equilibrium.* Liquid or gaseous phases interdiffuse and readily become homogeneous. Solid phases also interdiffuse, but because of the very limited mobility of atoms of most solids, the process is much slower. (See Chapter 5.) Therefore, the structures of almost all solids in everyday use are not in equilibrium.

There are two common uses of the word *state* which tend to become confusing and must now be distinguished. It is usual to refer to "the liquid state," "the solid state," and so on. In these cases, the references are to the *state of aggregation,* and not to the *thermodynamic state.* If the thermodynamic state of a system is determined, the states of aggregation of its contents are fixed. This is implicit in the definition. However, the reverse is not true, and care should be taken not to confuse the two terms. A system in a given thermodynamic state may contain only *one* phase in the gaseous state but one or more phases in the liquid state or in the solid state.

*Components* are those elements, compounds, or multiphase mixtures of which a system is composed. However, element A and element B may be mixed together to produce a total composition which also may be made from a mixture, for example, of A and a compound such as $A_3B$. The true set of components is considered to be any set which will do the job. If, for example, a 50% A–50% B composition is desired, the components $A_3B$ and A are obviously unsuitable since no mixture of the two will give a "50-50" composition. Many sets of components, however, may be considered to make up any given system, and the choice of one set or another is a matter of convenience. If a series of compositions is being considered, of course, components must be chosen which will make up *all* of the compositions.

The term *phase* also may not be clear at times. If the pressure or temperature of a system is changed, the compositions of the phases may change; if the total composition of the system is changed, the compositions of the phases may change. Should phases whose compositions are changed in such ways be regarded as new phases or not? Although the physical properties of the phases may be somewhat altered, the crystal structures will be similar. There is enough similarity, on the whole, so that the "new" set of phases may be considered to be a *continuation* of the "old" set. Therefore, a single phase may exist over a range of composition. As long as the change in physical properties caused by the change in composition is *continuous* and *smooth,* the same phase is present. If *all* the properties change *discontinuously,* the old phase is said to have been transformed to a new one. Changes in temperature, pressure, composition, or other variables may therefore merely change the properties of those phases that are present, may change the relative amounts of the phases, or may cause certain phases to appear or disappear. Because the properties of a system depend to a high degree on its *phase constitution,* the study of the *phase equilibrium* is extensive.

## 2.2 SINGLE-COMPONENT SYSTEMS: ALLOTROPY

Suppose that only pressure and temperature are considered in determining the equilibrium thermodynamic state of water. (This is a close approximation to reality in solids and liquids; other variables have a much smaller effect.) It is possible now to diagram the *equilibrium* phase constitution of water at all combinations of $P$ and $T$ which have been observed. Such a compilation of observations is called a *P-T Phase Diagram*, which is exemplified in Figure 2.1.

To use a diagram such as Figure 2.1, we locate points in *P-T* space by specifying values for the two variables. If such a point lies in the area labeled "liquid," that would be the state (and phase) in which $H_2O$ exists. If the point is on the line (e.g., *DC*), two phases would exist in *equilibrium*. If the specified temperature and pressure are the coordinates of point *D*, all three phases coexist in equilibrium with one another. Point *D* is called the *triple point*. It is located, for $H_2O$, at $0.0098°C$ and 4.58 mm of mercury pressure. Notice that the three-phase equilibrium is *invariant;* that is, it exists at only one point. For two phases to exist at equilibrium (as is the case along the lines *AD, DC,* and *BD* in Figure 2.1), if either $P$ or $T$ is varied, the other must assume a specific value. Therefore, in a one-component system, a two-phase equilibrium is *univariant.* If only one phase is to remain present, any combination of $P$ and $T$ which lies in the *phase field* of interest (i.e., in the solid, liquid, or vapor region) may be selected. This is a *bivariant equilibrium.* These rules apply to the *P-T* diagram of any one-component system. Two more familiar points are also shown in Figure 2.1. *E* represents liquid-solid equilibrium at one atmosphere pressure and $0°C$, and *F* represents liquid-vapor equilibrium at one atmosphere pressure and $100°C$.

A diagram such as Figure 2.1 may be used to describe the phase or phases present at a given set of $P$ and $T$ conditions. It may also be used to indicate at what point phase changes would occur when one quantity is varied and the other held constant. The slopes, directions, and intersections of the various lines in Figure 2.1 can be related to the thermodynamic properties of $H_2O$.

Since the study of materials is usually limited to the consideration of condensed systems (liquid and solid), it is important to consider the thermodynamic description of *allotropic transformations,* which are phase changes occurring in a single state of aggregation. These occur in many systems, such as $SiO_2$, $Al_2O_3$, Sn, Ti, Zr, and Fe. The allotropic transformations of iron are the most thoroughly understood of all and are of great practical

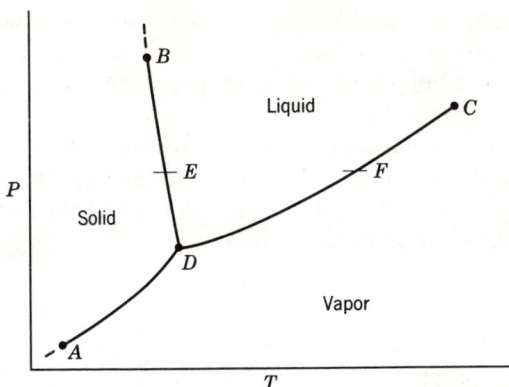

Figure 2.1    *P-T* phase diagram for $H_2O$.

importance, since they cause the wide property variations which may be produced by the heat treatment of steel.

Solid and liquid iron are normally handled at near-atmospheric pressure. At this pressure condensed-phase equilibria are relatively insensitive to small pressure variations. Accordingly, the phase diagram for pure iron may be drawn on a temperature coordinate only, as in Figure 2.2.

The free energy of the crystallographic form stable within a particular temperature range must be less than that of any other form (equation 1.20). From the experimental observation of phases present, and the temperatures at which they are present, and from a calculation of heat capacities and enthalpy changes accompanying various phase changes, the free energy of each of the allotropic forms of iron can be expressed as a function of temperature. To do this, the relations between *F*, *H*, *S*, and $C_p$ given in Chapter 1

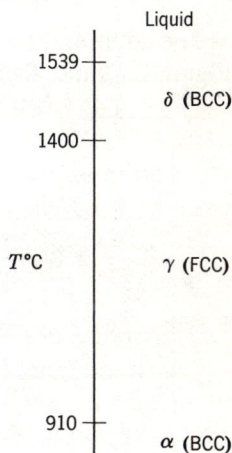

Figure 2.2    Phase diagram of pure iron at one atmosphere pressure.

must be employed. In principle, it is possible to calculate the free energy curves shown in Figure 2.3, and then by noting from the curves the range of stability of each phase, the actual phase diagram could be constructed.

A number of conclusions can be drawn from this figure. The general decrease in free energy of all the phases with increasing temperature is a result of the increasing dominance of the temperature-entropy term in the relation

$$F = H - TS \tag{2.1}$$

The increasingly negative slope for phases which are stable at increasingly higher temperatures is the result of the greater entropy of these phases, since (from equation 1.18)

$$\left[\frac{dF}{dT}\right]_P = -S \tag{2.2}$$

The greater curvature (smaller radius) of the curve for BCC iron than that for FCC iron is caused by the fact that BCC heat capacity is greater than that of FCC and because $C_P$ is always positive:

$$\left[\frac{d^2F}{dT^2}\right]_P = -\left[\frac{dS}{dT}\right]_P = -\frac{C_p}{T} \tag{2.3}$$

The addition of a second component alters the three curves in Figure 2.3, and their relative positions change. The allotropic

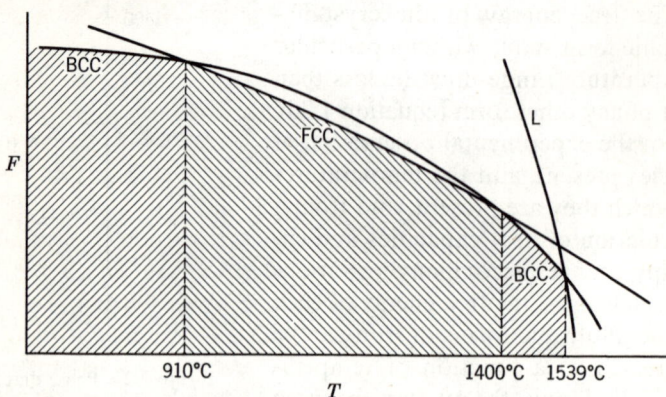

Figure 2.3    Free energy of iron phases as a function of temperature.

phase changes would therefore occur at different temperatures. The way in which the data in Figure 2.3 change with composition can be represented by the use of a composition coordinate axis, $X$, orthogonal to the $F$ and $T$ axes. In a binary (two-component system) only one additional axis is required, which can be located perpendicular to the $F$-$T$ plane to form a three-dimensional diagram with coordinates $F$, $T$, and $X$.

## 2.3    COMPOSITION AS A VARIABLE

The free energy concept may be used to analyze the influence of composition on phase constitution. Two-component systems will be primarily considered in this section, but extension to the many-component case will be indicated when necessary. The mixing of two miscible components into a single, homogeneous solution is in all cases an irreversible process. Once they are mixed, complete unmixing is impossible unless we have the aid of the legendary, microscopic "Maxwell Demon," who is reputed to have the ability to separate mixtures, atom by atom. Consequently, mixing always increases the entropy, and we would expect such an increase to be largest near the 50-50 composition, as depicted in Figure 2.4. The dashed line in Figure 2.4$a$ represents the entropy of the two components, at all compositions, before mixing. The increase in entropy due to mixing is then simply the difference between the ordinate of the total entropy curve in Figure 2.4$c$ and the dashed line. This quantity, the *entropy of mixing*, designated by the symbol $\bar{S}$, is represented in Figure 2.4$b$. Such an increase in entropy or, equivalently (equation 2.1), a decrease in free energy, signifies a tendency on the part of the two components to mix.

The other contribution to the free energy, the enthalpy, may signify either mixing or unmixing tendencies. In solids, volume changes are relatively small and, therefore, (equation 1.5) enthalpy may be approximated by the internal energy. A qualitative analysis of the enthalpy of mixing may then be derived by bonding energy considerations. If A atoms prefer to be in the vicinity of B atoms rather than other A atoms, and B atoms behave in a like manner, the energy (or enthalpy) will be lowered by mixing. This is true because bond energies are negative for stable bonds, with stronger

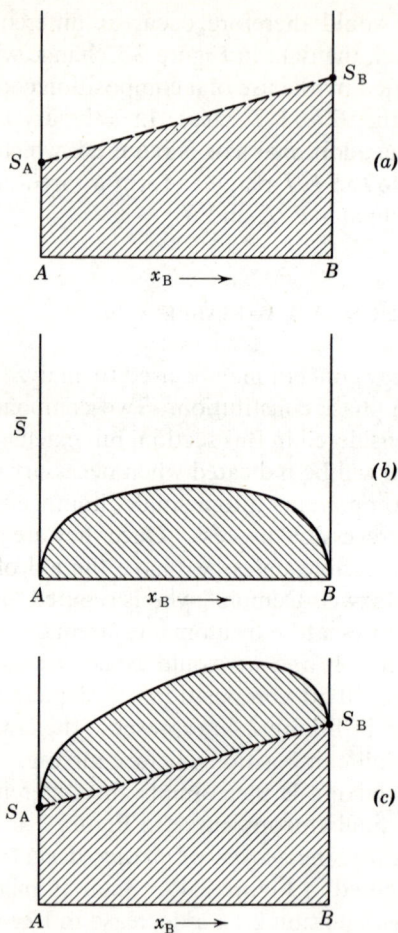

Figure 2.4  (a) Total entropy before mixing.  (b) Entropy increases due to mixing. (c) Total entropy of the mixture.

bonds more strongly negative.    Consequently, in Figure 2.5a, where the dashed lines are the total enthalpy *before* mixing, the enthalpy after mixing lies below the dashed line at all compositions. The maximum enthalpy decrease is expected to lie near the 50-50 composition, because a maximum number of "A-B bonds" would be formed there, if the mixture is truly homogeneous.    This situation is depicted in Figure 2.5*a*, and is sometimes referred to as a *negative deviation from ideality*. The ideal case is the middle diagram, Figure 2.5*b*. Here bonding strength is equal for A-A, B-B, and A-B combinations, and no enthalpy change occurs on mixing. A *positive deviation* is shown in Figure 2.5*c*. Such a situation occurs if A and B atoms prefer the company of their own kind. The average bond energy after mixing is less negative and therefore enthalpy is raised by mixing. These considerations are, of course, oversimplified. Lattice strains due to atomic size differences have not been considered. The actual energetics of such solutions are more complicated than single-atom bonding and many solid solutions are not completely homogeneous at equilibrium. Figure 2.5 is, nevertheless, qualitatively correct in almost all cases.

We may now obtain the free energy of a typical solution by multiplying the ordinate of Figure 2.4*c* by the temperature and subtracting this quantity from the enthalpy. Figures 2.6*a, b,* and *c* cover the three cases described in Figure 2.5. For negative deviations and ideal solutions, we see immediately that mixing decreases the free energy and conclude that such solutions are stable. The positive deviation is more interesting. At higher temperatures, the entropy term would dominate the free energy change on mixing, and the solution would always be stable. At lower temperatures (Figure 2.6*c*), the positive contribution from the enthalpy shows up. The free energy of mixing for solutions with intermediate compositions (i.e., near the middle of Figure 2.6*c*) may be positive, signifying that such solutions are unstable. To find the stable

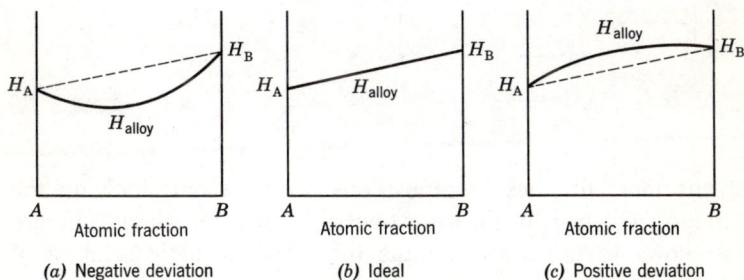

(a) Negative deviation        (b) Ideal        (c) Positive deviation

Figure 2.5    Enthalpy-composition diagrams.

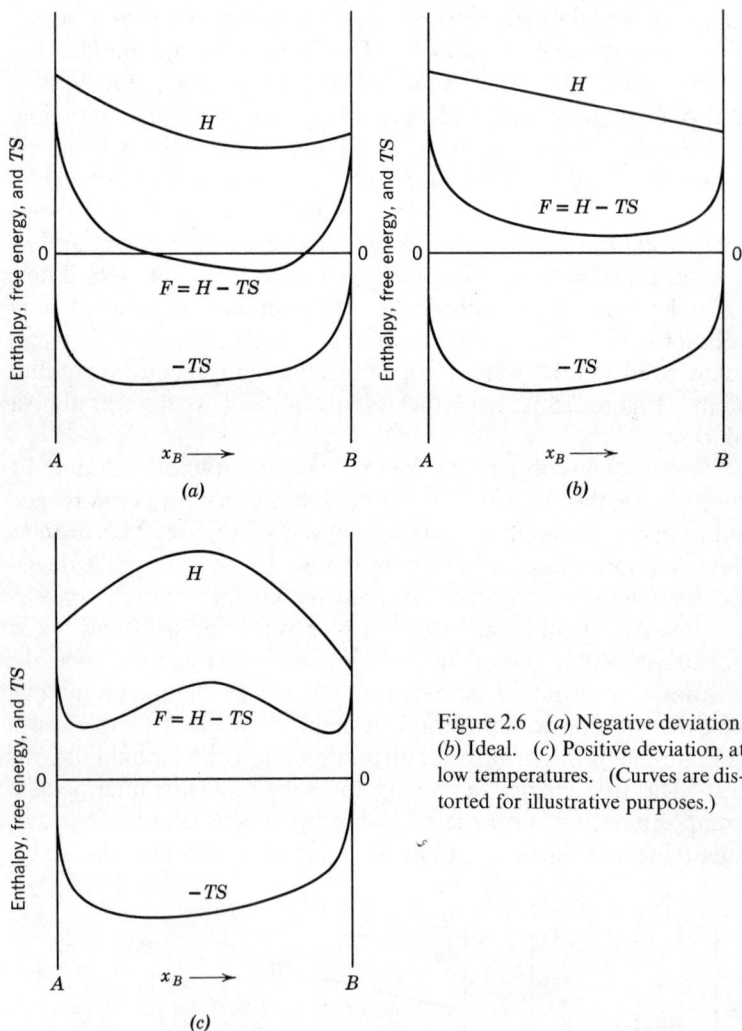

(a)

(b)

Figure 2.6   (a) Negative deviation. (b) Ideal.   (c) Positive deviation, at low temperatures.   (Curves are distorted for illustrative purposes.)

(c)

arrangement for these compositions, we need only look for the situation which yields the lowest possible total free energy.   Figure 2.7 shows what we are looking for.   The unstable solution of composition $x_3$ splits up into a mixture of two solutions of compositions $x_1$ and $x_2$ respectively.   The free energy of the mixture lies on the dashed line, for all total compositions between $x_1$ and $x_2$. (Proof of this will be given in Section 2.5.) Since the free energy of the mixture is less than that of the single solution over this entire composition range, we conclude that, for all total compositions between $x_1$ and $x_2$, a mixture of two solutions of those two compositions is the most stable state. Note that this is so even for compositions where the free energy of mixing is negative.

The procedure for finding stable phase equilibria may also be applied in binary systems where a number of alternate phases are formed. Consider Figure 2.8. The stable states for all compositions are described by the lowest possible "envelope" of free energy curves and tangent lines. In Figure 2.8, $\gamma$ and liquid are unstable at all compositions, $\alpha$ is stable for A-rich mixtures, and $\beta$ for B-rich mixtures. At overall compositions between the points of tangency with the dashed line, mixtures of $\alpha$ and $\beta$ are stable. The $\alpha$ and $\beta$ phases always have compositions corresponding to the points of tangency. Of course, the situation changes with changing temperature. If the temperature is high enough, the free energy-composition curve for the liquid will lie below all other curves and tangent lines, signifying stability of the liquid phase at all compositions. If the relative positions of the free energy-composition curves for a binary system are known for a series of temperatures, the stable phases or mixtures of phases may be deduced immediately for each temperature. These data may then be plotted in temperature-composition space; the result is the familiar *phase diagram,* which maps out the regions of stability for

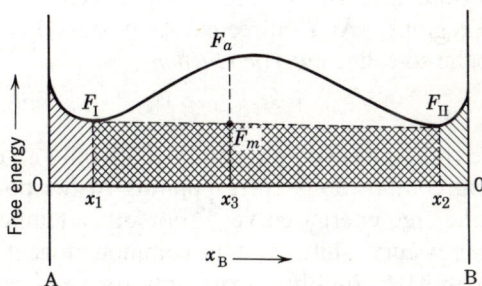

Figure 2.7  Free energy diagram for a binary alloy, showing separation into two terminal solid solutions.

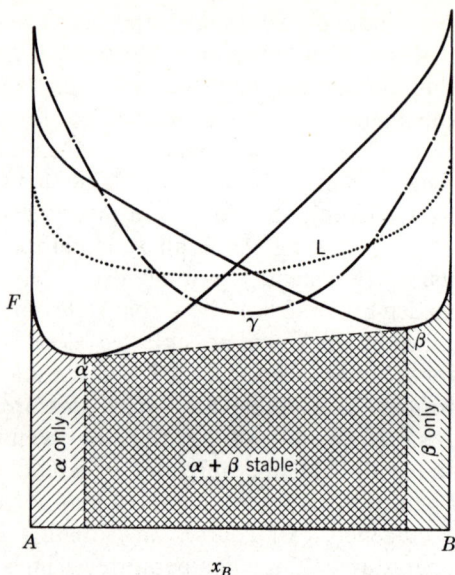

Figure 2.8   Hypothetical binary system showing two terminal solid solution phases $\alpha$ and $\beta$, an intermediate phase $\gamma$, and a liquid.

each phase and phase mixture.   Figure 2.9 depicts the development of a eutectic phase diagram from a series of free energy–composition diagrams.   At each temperature, the appropriate tangents are drawn to the free energy curves.   The points of tangency mark the constant compositions of the two phases in the two-phase (mixture) region, and the boundaries between two-phase and single-phase regions.   At $T_2$ three curves possess a common tangent.   Here, the so-called *eutectic reaction,*

$$L \rightleftharpoons \alpha + \beta \qquad (2.4)$$

occurs.   The three-phase equilibrium occurs only at $T_2$, between phases having compositions corresponding to the points of tangency with the free energy curves.   For other temperatures the liquid free energy curve shifts, and the common tangent disappears. Because three-phase equilibria exist only for fixed temperatures and phase compositions in binary systems they are called *invariant.*
From the various free energy–composition diagrams, we may

"predict" phase diagrams for typical binary systems.  For a system having a single ideal or negatively deviated solid solution (Figures 2.6b and a) only one solid phase is present.  As the temperature rises, the free energy curve for the liquid drops relative to

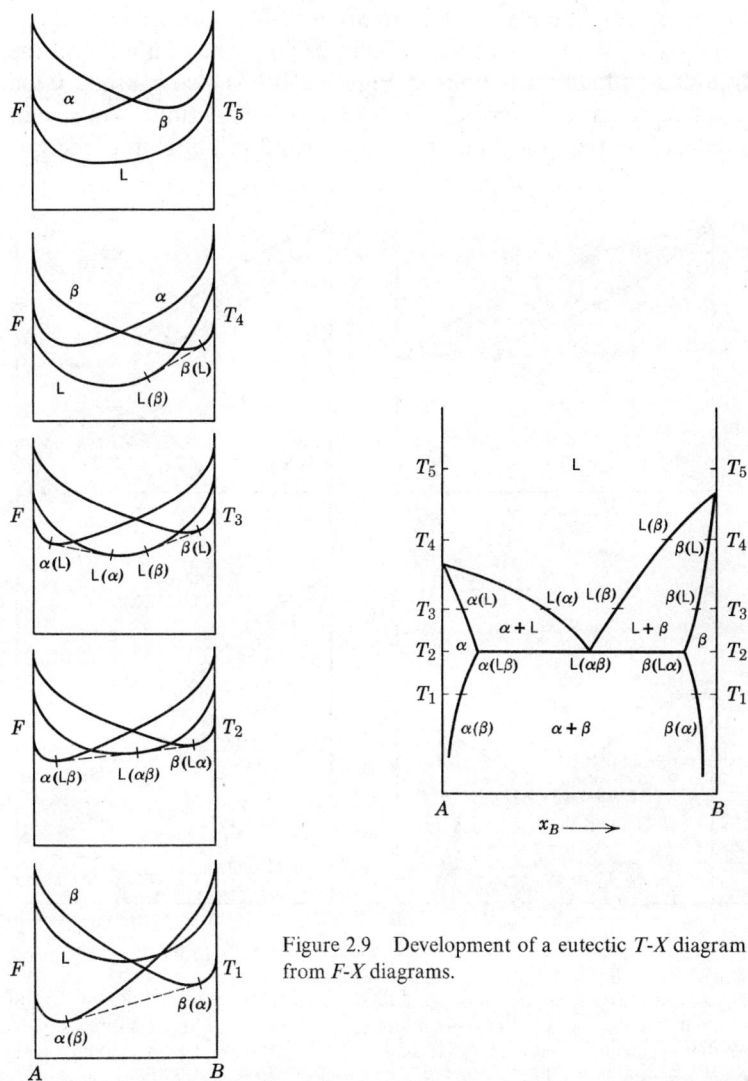

Figure 2.9    Development of a eutectic T-X diagram from F-X diagrams.

that of the solid, and a situation similar to that at $T_4$ in Figure 2.9 develops. The final diagram will resemble Figure 2.10a. If the solid solution has a positive deviation (Figure 2.6c), we expect it to behave as described previously in this section: complete solubility at high temperatures, breaking up into two stable solutions at lower temperatures. This situation, called a *miscibility gap*, is depicted in the lower part of Figure 2.10b. The minimum in the liquidus and solidus curves in Figure 2.10b is also caused by the positive deviation from ideality in the solid solution. We saw in Figure 2.6c that the "hump" in the solid-solution free energy–

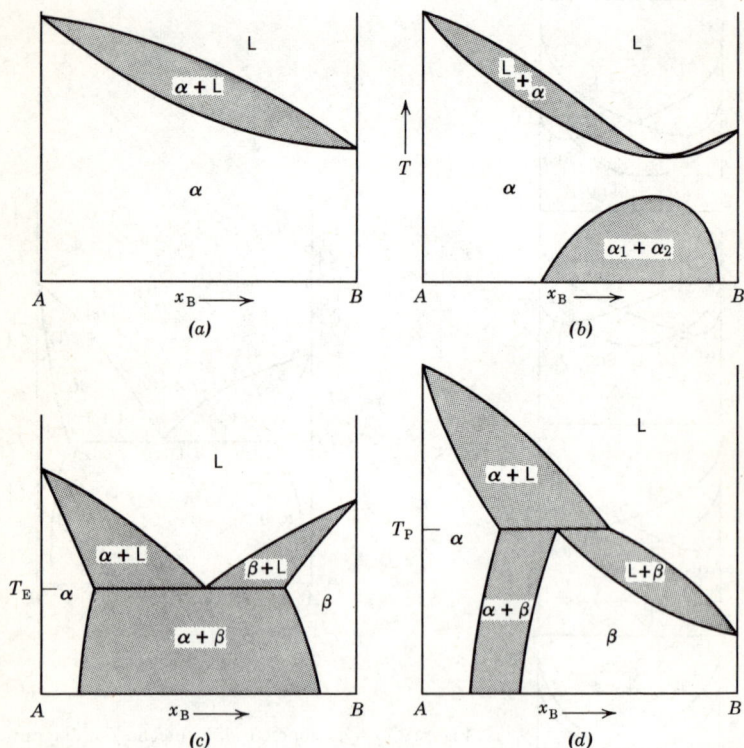

Figure 2.10   (a) Complete solid solubility.   (b) Melting point minimum.   Solid solubility almost complete for "miscibility gap": $\alpha \rightleftharpoons \alpha_1 + \alpha_2$.   (c) Limited solid solubility, eutectic reaction at $T_E$: $L \rightleftharpoons \alpha + \beta$.   (d) Limited solid solubility, peritectic reaction at $T_P$: $L + \alpha \rightleftharpoons \beta$.

composition curve exists only at low temperatures. As the temperature rises the "hump" eventually flattens out. A complete series of solid solutions forms only if the liquid free energy curve does not cut into the "hump" before it flattens out. In this event the solid-solution free energy curve is generally less sharply curved than that of the liquid, and a *minimum melting point* occurs as in Figure 2.10*b*. On the other hand, if the liquid free energy curve descends and intersects that of the solid solution before its "hump" flattens out, a eutectic reaction results. The phase diagram produced is similar to Figure 2.10*c* except that both terminal solid solutions have the same crystal structure.

In general the eutectic reaction shown in Figure 2.10*c* results from the free energy curve intersections and tangents discussed for Figure 2.9. Figure 2.10*d* occurs many times if one component has a relatively low melting point. The free energy curve for the liquid descends first to the right of the $\beta$ curve rather than the left as in Figure 2.9 as the temperature rises. The invariant three-phase transformation,

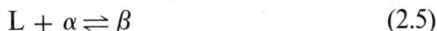

$$L + \alpha \rightleftharpoons \beta \qquad (2.5)$$

is called a *peritectic reaction.*

Phase diagrams may, of course, contain more than one invariant transformation. Figure 2.11 combines the eutectic

$$L \rightleftharpoons \sigma + \beta$$

the peritectic

$$\alpha + L \rightleftharpoons \sigma$$

and the eutectoid

$$\sigma \rightleftharpoons \alpha + \beta \qquad (2.6)$$

## 2.4   RULES OF SOLID SOLUBILITY

To some extent positive deviations of the enthalpy of mixing, and consequent limited solid solubility, may be predicted from known atomic properties. In the past two decades W. Hume-Rothery and his colleagues have set up empirical rules to predict

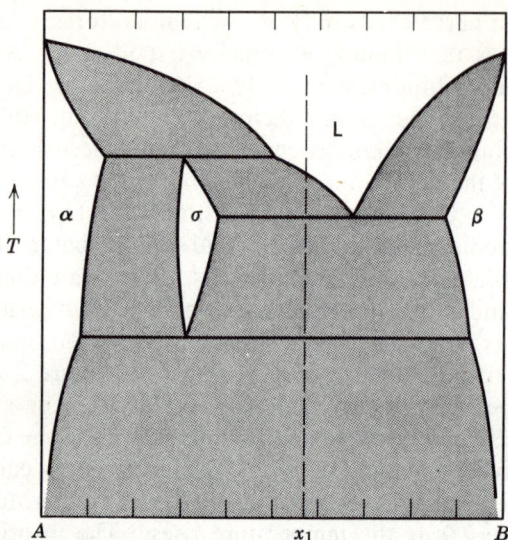

Figure 2.11   A more complicated phase diagram.

the existence of extensive solid solubility.   For this condition to exist between two elements, the following statements must, almost always, be true:

1. *Atomic size:* the atomic radii of the two elements must be within 15% of each other.

2. *Crystal structure:* the type of crystal structure must be the same.

3. *Chemical valence:* the valence of the two elements must differ by no more than one.

4. *Electronegativity:* the electronegativities must be nearly equal; if not, a compound may be formed as a result of the difference in affinity for electrons.

The principle of the fourth rule is that the single solution becomes unstable with regard to the compound, whose free energy near its ideal composition will be lowest.   Two-phase regions (compound plus solid solution) then appear, and the phase diagram no longer contains the broad region of complete solubility.   If electron affinity is the same for each component, no compound will be formed, and the one-phase region will be retained. The first three rules are based on the increase in enthalpy due to distortion of the crystal lattices, disruption of crystal structure, and, in a very crude sense, unsaturated bonds. All of these may lead to appreciable positive deviations and, therefore, limited mutual solubility. If some, but not all the rules are obeyed, relatively large areas of the phase diagram, separated by two-phase gaps, may show some range of solubility.

We might expect such energy changes to affect the relative stabilities of the phases present at different temperatures. There are, however, complications, and no set of rules as consistent as those of Hume-Rothery appear to exist. Some regularities may be noted. The addition of nickel (FCC) to iron increases the range of stability of the FCC solid solution at the expense of the BCC form, as shown in Figure 2.12a. Conversely, addition of chromium (BCC) expands the BCC range, and shrinks that of the FCC. Similar effects occur for certain additions to titanium and zirconium. Such behavior is not, however, regular enough to justify generalization.

## 2.5   THE LEVER RULE

Having developed the idea of the phase diagram by free energy techniques, we should not forget that phase diagrams are direct records of experimental evidence. We do not, in general, measure free energies in order to determine phase constitution. It is usually much easier to allow samples of known composition to come to equilibrium at various temperatures and to search, by use of optical, electrical, or x-ray techniques for the presence of the various phases. These data can be recorded directly on the diagram. The phase diagram is also the most useful tool for the prediction of phase constitution. The remainder of this chapter will be devoted to the extraction of information from phase diagrams.

Consider the alloy system of Figure 2.13. If we start with pure A at temperature $T_1$ and slowly add B, our entire specimen will be $\alpha$ phase until the composition becomes $x_\alpha$. As the B fraction in the alloy exceeds $x_\alpha$, $\beta$ phase will appear. Once in the two-phase region, the compositions of $\alpha$ and $\beta$ at temperature $T_1$ remain

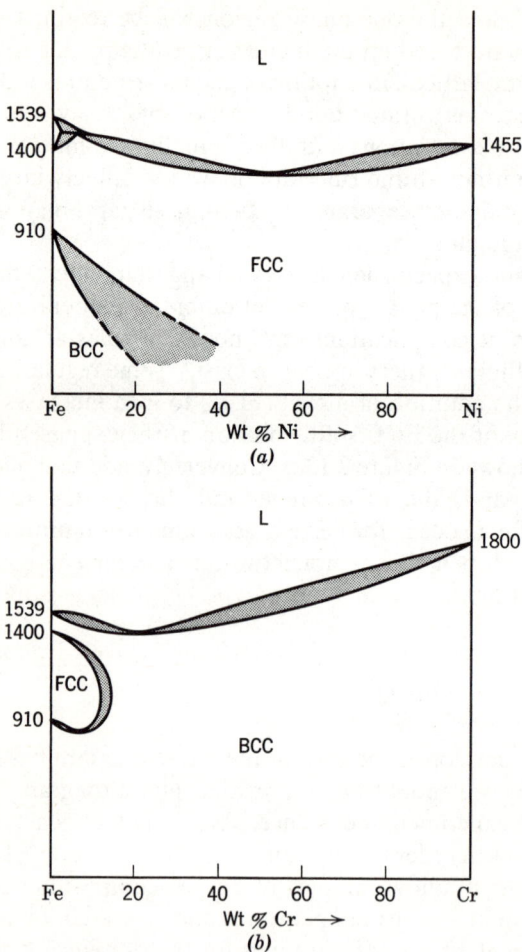

Figure 2.12   Phase diagrams for iron with additions of (a) nickel and (b) chromium.

fixed and further increases in total B content occur by increases in the amount of $\beta$ phase present, and decreases in $\alpha$ phase. This goes on until the total composition reaches $x_\beta$, at which point the alloy is totally $\beta$. Further enrichment of the B content of the alloy merely increases that of the $\beta$ phase. Because the compositions of $\alpha$ and $\beta$ are both fixed for the two-phase alloy, the relative

amounts of each phase may be computed for any total composi-
tion between $x_\alpha$ and $x_\beta$. Material conservation is used. If $N_\alpha$
moles of $\alpha$ and $N_\beta$ moles of $\beta$ are present,

$$N_\alpha + N_\beta = N \qquad (2.7)$$

where $N$ is the total number of moles of alloy. Likewise, the
total moles of B in the alloy must equal the sum of B in both phases:

$$x_\alpha N_\alpha + x_\beta N_\beta = x_0 N \qquad (2.8)$$

Solution of the above pair of equations for $N_\alpha$ and $N_\beta$ gives

$$\frac{N_\alpha}{N} = \frac{x_\beta - x_0}{x_\beta - x_\alpha}; \frac{N_\beta}{N} = \frac{x_0 - x_\alpha}{x_\beta - x_\alpha} \qquad (2.9)$$

for the fraction of total material which exists as either phase.
Obviously the above applies to weight fraction also. If the differences $x_\beta - x_0$ and $x_0 - x_\alpha$ are regarded as *lever arms,* the reason
for the term *lever rule* becomes apparent. The total composition
$x_0$ is the "fulcrum" and the amount of each phase is proportional
to the length (i.e., composition difference), of the opposite lever
arm. The fraction of the alloy which comprises each phase is
simply the opposite lever arm divided by the total "length",
$x_\beta - x_\alpha$.

The total free energy of the alloy is

$$NF_0 = N_\alpha F_\alpha + N_\beta F_\beta \qquad (2.10)$$

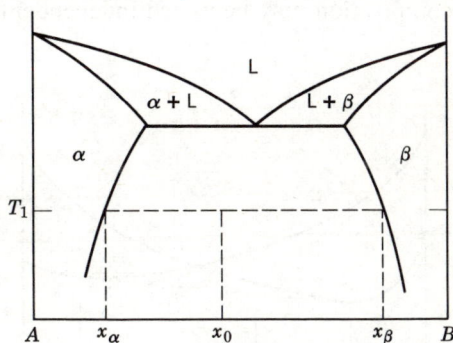

Figure 2.13   The phase diagram for the system *A-B,* and application of the Lever
Rule to the two-phase region.

where $F_0$, $F_\alpha$, and $F_\beta$ are the free energies per mole of the alloy and the phases, respectively, when two phases are present. Using equation 2.9 with the above results in

$$F_0 = \left\{\frac{x_\beta - x_0}{x_\beta - x_\alpha}\right\} F_\alpha + \left\{\frac{x_0 - x_\alpha}{x_\beta - x_\alpha}\right\} F_\beta \qquad (2.11)$$

for a mole of alloy. Rearranging,

$$F_0 = F_\alpha - \left(\frac{x_0 - x_\alpha}{x_\beta - x_\alpha}\right)(F_\beta - F_\alpha) \qquad (2.11a)$$

of the straight line connecting the points $(F_\alpha, x_\alpha)$ and $(F_\beta, x_\beta)$, and is also the free energy of the mixture of phases. The lowest possible straight line, where it lies below all other free energy curves, corresponds to the lowest free energy and therefore the equilibrium state of the system. The tangent line (dashed) in Figure 2.14 is such a line.

## 2.6   THE PHASE RULE: FREE ENERGY RECONSIDERED

We have observed in the binary systems of the previous sections that if two phases are present at constant pressure and some known temperature, the compositions of both phases are fixed. If the temperature is changed, the equilibrium compositions of the phases in the two-phase field also change. By contrast, both temperature and composition may be varied independently when one

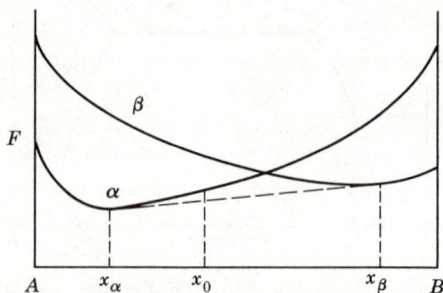

Figure 2.14   The Lever Rule for free energy-composition diagrams.

phase is present. Three phases exist in equilibrium only on the horizontal lines corresponding to invariant reactions (i.e., at fixed temperature and composition). It seems then, that as the number of phases present increases, the number of independent variables decreases; similar behavior was seen for the one-component system in Section 2.2. We shall now show that this behavior is universal, and stems from the free energy criterion for equilibrium.

If $M$ variables are related by $N$ equations, there are only $M - N$ truly independent variables; any $N$ variables may be eliminated by appropriate solution of the equations. If $M = N$, a complete and unique solution is possible. If, for instance, three variables are related by two equations, two of the variables may be expressed in terms of the third, and their values determined by assigning a value to the third. The thermodynamic variables in the present case are temperature, pressure, and composition of each phase. If the system has $C$ components and there are $P$ phases present, there are $(CP)$ composition variables, $C$ for each phase. The total number of variables is therefore $CP + 2$.

The first set of equations relating the above variables is the requirement that the sum of the mole fractions of all the components in any phase adds up to unity. There are $P$ such equations, one for each phase. The other set of equations comes from the free energy criterion for chemical equilibrium. Consider the possibility of a very small quantity of any component being transferred from one phase to another. At equilibrium, the resulting change in free energy must be zero. If, for instance, the free energy change is negative, equilibrium has not been achieved, the component will spontaneously transfer, and compositions change until free energy is minimized. If, on the other hand, the free energy change is positive, the transfer of material in the opposite direction will reduce free energy, and the above argument applies again.

The total free energy change occurring during the transfer discussed above is the sum of the changes which occur in each of the two phases. If $N_j{}^i$ refers to the number of moles of component $j$ in phase $i$, and $F_i$ the free energy of phase $i$, the change in free energy corresponding to the removal of a small amount, $\delta N_j{}^i$, of component $j$ from the phase is, to the first order,

$$\delta F_i = \frac{\partial F_i}{\partial N_j{}^i} \delta N_j{}^i \qquad (2.12a)$$

If the material is now deposited in phase $k$,

$$\delta F_k = \frac{\partial F_k}{\partial N_j{}^k}\, \delta N_j{}^k \tag{2.12b}$$

But $\delta N_j{}^k = -\delta N_j{}^i$, since the same amount of component $j$ was removed from phase $i$ and added to phase $k$. Consequently, if the total free energy change due to the transfer is to be zero,

$$\delta F_{\text{(total)}} = \delta F_i + \delta F_k$$

$$= \left[ \frac{\partial F_i}{\partial N_j{}^i} - \frac{\partial F_k}{\partial N_j{}^k} \right] \delta N_j{}^{i\ \text{or}\ k} = 0 \tag{2.13a}$$

The general form, for multiphase chemical equilibrium is then

$$\frac{\partial F_i}{\partial N_j{}^i} = \frac{\partial F_k}{\partial N_j{}^k} \tag{2.13b}$$

*for any two phases,* which means that the quantity $\partial F_i / \partial N_j{}^i$, must be equal for all the phases present. The complete equations for equilibrium are

$$\frac{\partial F_1}{\partial N_{\mathrm{I}}{}^1} = \frac{\partial F_2}{\partial N_{\mathrm{I}}{}^2} = \frac{\partial F_3}{\partial N_{\mathrm{I}}{}^3} = \cdots \frac{\partial F_P}{\partial N_{\mathrm{I}}{}^P}$$

$$\frac{\partial F_1}{\partial N_{\mathrm{II}}{}^1} = \frac{\partial F_2}{\partial N_{\mathrm{II}}{}^2} = \cdots$$

$$\frac{\partial F_1}{\partial N_{\mathrm{III}}{}^1} = \cdots$$

$$\frac{\partial F_1}{\partial N_C{}^1} = \cdots\cdots\cdots = \frac{\partial F_P}{\partial N_C{}^P} \tag{2.14}$$

for any possible material transfer of the $C$ components between the $P$ phases. Each of the $C$ rows of equation 2.14 contains $(P - 1)$ equalities. Therefore, the free energy criterion for equilibrium imposes $C(P - 1)$ equations. Summing up: our hypothetical $C$-component, $P$-phase system has $CP + 2$ variables, linked by two sets of equations. One set has $P$ equations, the other $C(P - 1)$. The number of independent variables is then

$$V = CP + 2 - P - C(P - 1)$$

$$= C - P + 2 \tag{2.15}$$

The phase diagrams we have discussed refer to systems at one atmosphere pressure (i.e., pressure is voluntarily fixed, removing one independent variable):

$$V_{p=\text{const.}} = C - P + 1 \qquad (2.15a)$$

For a binary system, $C = 2$ and $V = 3 - P$. Thus, three-phase equilibria are invariant, two-phase equilibria have only one (temperature *or* composition) variable, and one-phase equilibria have two.

## DEFINITIONS

*Allotropic forms.*  Alternative arrangements of atoms in an element or compound.

*Component.*  A raw material used to form a multiphase system.

*Electronegativity.*  A numerical system describing the relative tendencies of atoms to acquire electrons.

*Enthalpy of Mixing ($\bar{H}$).*  The net change in enthalpy due to the mixing of components.

*Entropy of Mixing ($\bar{S}$).*  The net change in entropy due to the mixing of components.

*Eutectic Reaction.*  $L \rightarrow \alpha + \beta$ (i.e., two solid phases) on decreasing temperature.

*Eutectoid Reaction.*  $\sigma \rightarrow \alpha + \beta$ (all solid phases) on decreasing temperature.

*Ideal Solution.*  Zero enthalpy of mixing.

*Negative Deviation.*  Negative enthalpy of mixing.

*Peritectic Reaction.*  $\alpha + L \rightarrow \sigma$ on decreasing temperature.

*Phase.*  A physically distinct region of matter having characteristic atomic structure and properties which change continuously with temperature, composition, and any other thermodynamic variable.  The various phases of a system are, in principle, mechanically separable.

*Solid Solubility.*  The irreversible mixing of two solids into a single phase.

*State of Aggregation.*  Gaseous, liquid or solid.

## BIBLIOGRAPHY

Cottrell, A. H., *Theoretical Structural Metallurgy*, St. Martin's, New York, 1957. Pages 102 to 105 cover the entropy of mixing of crystals.  Chapters 10 and 11 deal with free energy of alloy phases and phase equilibria.  Undergraduate level.

Darken, L. S., and R. W. Gurry, *Physical Chemistry of Metals*, McGraw-Hill, New York, 1953.  Chapters 10, 11, and 12 cover phases and phase equilibria. Chapter 10 is a detailed treatment of solutions and mixing.  Graduate level.

Guy, A. G., *Elements of Physical Metallurgy*, Addison-Wesley, Reading, Massachusetts, 1951, second edition, 1959.   Chapter 6 deals with phase diagrams from the experimental point of view; excellent photomicrographs.   Detailed account of iron-carbon system, aluminum-silicon, and various magnesium alloys. Undergraduate.

Hansen, M., and K. Anderko, *Constitution of Binary Alloys*, McGraw-Hill 1958.   A comprehensive atlas, many inches thick, of all the metallic binary diagrams known at the time.

Kingery, W. D., *Introduction to Ceramics*, Wiley, 1960, Chapter 9 ("Ceramic Phase Equilibrium Diagrams").   Undergraduate level.

Levin, E. M., H. F. McCurdie, and F. P. Hall, *Phase Diagrams for Ceramists*, American Ceramic Society, Columbus, Ohio, 1956.   Ceramic phase diagrams.

*Metals Handbook*, American Society for Metals, Cleveland, 1948.   Many common binary and ternary diagrams, with other useful data.

## PROBLEMS

2.1   Refer to the *F-T* diagram, Figure 2.3.   Does the free energy *F always* decrease with increasing temperature?   Is the second derivative always negative?   (That is, are the curves always convex upwards?) Explain your answers.

2.2   Suppose four distinct phases were observed in a laboratory specimen of a binary alloy.   Is such an observation possible?   Explain your answer.

2.3   Figure 2.2 shows that the BCC ($\alpha$) form of iron is stable up to 910°C.   Above 910°, the FCC ($\gamma$) structure is stable.   If tungsten, molybdenum, or chromium is added to iron, the $\alpha$-$\gamma$ transformation temperature is raised.   However, if nickel is added, the $\alpha$-$\gamma$ transformation temperature is lowered.   Explain.

2.4   Consider a binary metallic solid solution.   Suppose the atomic forces were so short in range that the bonding energy depended only on nearest neighbor atoms.   Then the bonds between each atom and the $Z$ atoms coordinated about it ($Z$ is the coordination number, as in Volume I of this series) may be counted, and the total energy of bonding determined.   For a hypothetical A-B solution, $E_{AA}$ is the bond energy between two A atoms, $E_{BB}$ the same for two B atoms, and $E_{AB}$ the energy of an A-B bond.   Also in solids, the change in enthalpy is usually approximately equal to the change in energy, as $\Delta(PV)$ is quite small compared to $\Delta E$.   Show that the enthalpy of mixing for a mole of such a hypothetical binary solution is

$$\bar{H} = X_A X_B Z N_0 \left[ E_{AB} - \frac{E_{AA} + E_{BB}}{2} \right]$$

$N_0$ is Avogadro's number, that is, the number of atoms in a mole. Be careful not to count bonds twice; for instance, there are only $\frac{1}{2}ZN_0$ bonds in a mole of material.

2.5   Consider the result of Problem 2.4.   If, as we mentioned in Section 2.3, A-B bonds are preferred, then

$$E_{AB} - \frac{E_{AA} + E_{BB}}{2} < 0$$

since the $E_{AB}$ term is more strongly negative. Then

$$\bar{H} = X_A X_B \Omega \qquad \left(\Omega = ZN_0\left[E_{AB} - \frac{E_{AA} + E_{BB}}{2}\right]\right)$$

where $\Omega$ is a negative number.   Plot the quantity.

$$H_{\text{alloy}} = X_A H_A + X_B H_B + \bar{H}$$

on the same axes as those of Figure 2.5.   Use the values $H_A = 3$, $H_B = 5$, and $\Omega = -4$.   Compare with the diagram at the left of Figure 2.5 (i.e., negative deviation).   Do the same for positive deviation, that is, $\Omega > 0$. (Use $\Omega = +4$ for the calculation.)   Do the same for the ideal solution. What relation between $E_{AA}$, $E_{BB}$, and $E_{AB}$ must apply to ideal solution?

2.6   Component A is dissolved, in different concentrations, in each of two solutions.   Over each solution is gaseous A which is in equilibrium with the dissolved A.   Consequently, the pressure of A over each solution depends on the concentration of A in the solution.   There is a very large amount of each solution, and removal of a mole of A will not alter the concentration of dissolved A appreciably.   Suppose the following operations are now performed isothermally:

(a) Reversible evaporation of A from Solution 1 at its equilibrium partial pressure $P_1$.

(b) Reversible compression or expansion from Pressure $P_1$ to Pressure $P_2$, which is the equilibrium partial pressure for solution 2.

(c) Reversible dissolution of the gaseous A at pressure $P_2$ into solution 2.

The steps a, b, and c have transferred one mole of A from solution 1 to solution 2.   Find the free energy change $\Delta F$ for the transfer in terms of $P_1$ and $P_2$.   Use the result of Problem 1.11 for step b.   Does $\Delta F$ depend on the method of transferral?

2.7   If solutions 1 and 2 in Problem 2.6 are "ideal" solutions, then

$$P_1 = P_0 X_1, \qquad P_2 = P_0 X_2$$

where $P_0$ is the equilibrium vapor pressure of pure A, and $X_1$ and $X_2$ are the mole fractions of A in solution 1 and 2, respectively.   Find $\Delta F$ for the transfer in terms of the mole fractions. Suppose one mole of pure A were dissolved in solution 2. What is $\Delta F$ now? Use the Gibbs-Helmholtz equation (Problem 1.6) to find $\Delta S$.

2.8    Use the results of Problem 2.7 to find the entropy of mixing $\bar{S}$ of an ideal solution as in Figure 2.4. Proceed in the following manner: build up a binary solution of components A and B by adding each component such a way that their concentrations in the solution are constant. Prove that

$$\bar{S} = -R[X_A \ln X_A + X_B \ln X_B]$$

for one mole of solution, where $X_A$ and $X_B$ are the mole fractions of A and B, respectively. Plot $S$ versus composition as in Figure 2.4. If $S_A$ and $S_B$, the entropies per mole of components A and B are known, what is the total entropy of the solution? Compare your result with Figure 2.4. Find $\bar{F}$ for the solution, in the same way $\bar{S}$ was found, using the results of Problem 2.7. What is $H$? If $H_A$ and $H_B$, the enthalpies per mole of the pure components, are known, find the total enthalpy of the solution. Does your result agree with Figure 2.5?

2.9    Consider one mole of binary solid solution containing $N_A$ atoms of component A and $N_B$ atoms of component B. In the crystal structure of the solid solution there are, therefore, $N_A + N_B$ sites for the atoms. There are many ways of arranging the atoms on the sites. In fact there are

$$W = \frac{(N_A + N_B)!}{N_A! N_B!}$$

alternative arrangements. Statistical mechanics (which is discussed in most of the references in the bibliography of Chapter 1) shows that the increase in entropy due to the existence of such alternatives is

$$S = k \ln W.$$

From these statements derive the entropy of mixing $\bar{S}$ for the solution. Use the *Stirling Approximation,*

$$\ln (N!) = N \ln N - N$$

for very large $N$, to eliminate factorial terms. Compare your result with that of Problem 2.8, which was based on totally different assumptions.

2.10    Refer to Figure 2.11. Label the two-phase regions. Explain your answer. Suppose a mixture of composition $X_1$ were cooled slowly, from the liquid state. Describe the changes which occur on cooling, the new phases which appear, the transformations which occur, and the increase or decrease of the amounts of the phases present.

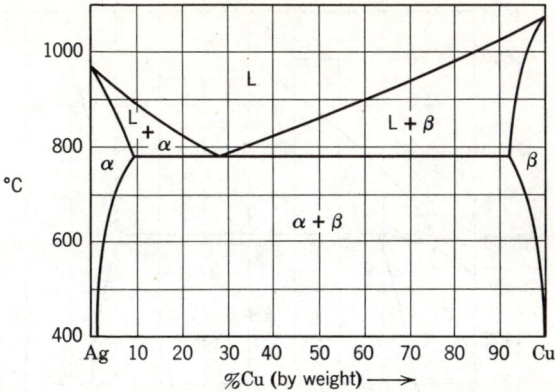

Figure 2.15   The silver-copper phase diagram.

2.11   In the silver-copper alloy system (Figure 2.15) the eutectic reaction

$$L \rightarrow \alpha + \beta$$

occurs on cooling.   The structure which commonly results from the eutectic reaction consists of fine, alternating layers of $\alpha$ and $\beta$.   Why? What are the compositions of the phases involved in the eutectic reaction? Suppose an alloy of composition 60% Cu is cooled from 1000°C.   Using the lever rule, calculate approximately:

(a) The fraction of the alloy (by weight) which exists as $\beta$ phase at 900°C.

(b) Similarly, the $\beta$ fraction present just above the eutectic temperature.

Draw a picture of the microstructure of the alloy just above the eutectic temperature.   Indicate the domains of the various phases.

(c) What fraction of the alloy is liquid, just above the eutectic temperature?

(d) Calculate the fractions of $\alpha$ and $\beta$ present, just below the eutectic temperature.

(e) What fraction of the total weight of the alloy exists as the layered eutectic structure, just below the eutectic temperature?

Draw the microstructure just below the eutectic temperature.   Finally, write a qualitative account of the events which occur on the cooling of the alloy, changes in amount and composition of the various phases, and invariant reactions.

2.12   Figure 2.16 is a high-temperature portion of the titanium-tungsten phase diagram.   Answer approximately, using the lever rule:

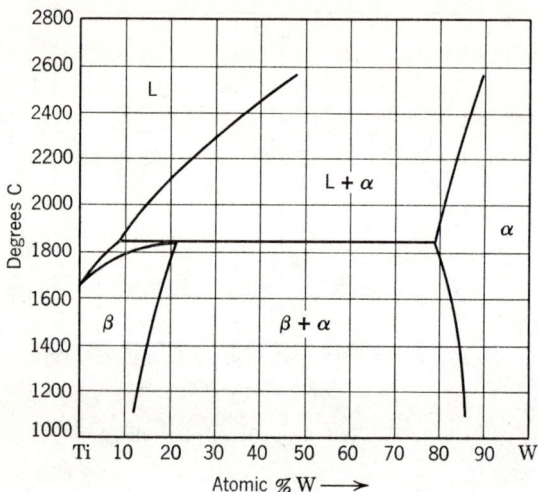

Figure 2.16   High-temperature phase diagram for the titanium-tungsten system.

(a) What atomic fraction of a 30% W alloy is solid $\alpha$ at 2300°C?

(b) What atomic fraction is $\alpha$ just above the peritectic temperature?

(c) Just below the peritectic temperature? What event is responsible for the change?

(d) What are the compositions of the phases which participate in the peritectic reaction?

(e) How much $\alpha$ and how much liquid are consumed to form one mole of $\beta$ by the peritectic reaction?

(f) Use your answer to (e) to check your answers to (b) and (c).

2.13   Show with the aid of a sketch the equilibrium state between two beakers, one containing a silver electrode and the other a platinum electrode. The second beaker contains $Ag^+$ ions in solution and the former $Fe^{++}$ and $Fe^{+++}$ ions. The two liquids are connected by a siphon and the two electrodes by a platinum wire.

(a) Write a chemical equation for the spontaneous change in the reversible chemical reaction in the cell.

(b) Now superimpose a voltage of $>0.28$ volts so that electrons flow from the Ag electrode to the Pt. Write the equation for the cell reaction.

2.14   If the solubility of calcium sulfate in water at 25°C is $6.2 \times 10^{-3}$ mole of $CaSO_4$ per liter, write

(a) The chemical equation of the reaction.

(b) Calculate the equilibrium constant.

(c) Calculate the calcium ion concentration if $[SO_4^{--}] = 0.200M$, and the solution is in equilibrium with solid calcium sulfate at $25°C$.

2.15 (a) What is meant by "coring" of a solid solution?

(b) How can it be removed in an aluminum alloy casting containing 4 per cent copper?

2.16 Why should it be difficult to grow single crystals of an alloy which undergoes a peritectic reaction?

2.17 Sketch a zone-refining apparatus and describe what kinds of alloys are most readily refined by this method. Illustrate your answers with suitable diagrams.

2.18 Describe the experimental apparatus used to determine phases and phase boundaries in binary alloy diagrams. Include thermal, electric resistance, x-ray diffraction, and optical methods.

CHAPTER THREE

# *Thermodynamics of Surfaces*

SUMMARY

The surfaces of phases always differ in behavior from the bulk of the phases themselves, because of the rapid structural changes which must occur at and near phase boundaries. Equilibrium bonding arrangements are disrupted, leading to an excess energy associated with the surface (i.e., *surface energy,* $\gamma$), which is measured in energy per unit area of the surface. The excess energy may be minimized by minimizing surface area. This tendency is called *surface tension.* Surface energy may also be lowered by segregation of the various components to and from the surface; such behavior is called *adsorption.* The magnitude of $\gamma$ may be estimated for metallic and covalent materials by considering the number and energy of the bonds which must be broken to form the surface. Similarly, calculations of work done against the coulomb force lead to approximate values of $\gamma$ for ionic materials. In both cases, $\gamma$ depends on crystallographic orientation. Direct measurement of $\gamma$ is possible by force equilibrium, if the phases are sufficiently mobile. The extent of adsorption which may occur may be calculated from known properties by thermodynamic considerations.

## 3.1 INTRODUCTION

A vital feature in every multicomponent system was neglected in Chapter 2 but will be discussed in this chapter. Having examined the nature of the phases themselves, we must now turn to the boundaries, commonly called *surfaces,* which separate the phases. At the interface between two phases, the crystal structure, or the state of aggregation, or the composition must change in a fairly abrupt manner. The atoms in the vicinity of the surface are not in equilibrium states, since they are in neither one phase nor the other. Unsatisfied bonds abound. The excess energy due to the

46

perturbed material at the surface is proportional to the surface area. Thus, a drop of liquid will tend to assume a spherical shape in order to minimize its surface area and, thereby, its surface energy. In single-phase solids, similar surfaces exist. These are the grain boundaries, which exist between grains of different crystallographic orientations.

Surfaces are important to the study of microstructures, friction and wear, the joining of all materials by all means, the catalysis of chemical reactions, oxidation, corrosion, the mechanical behavior of small or thin bodies, the design of electronic devices, and a wide variety of other phenomena. In this volume, the properties of surfaces will be discussed in many of the remaining chapters.

## 3.2   CHARACTERISTICS OF SURFACES

The description of surfaces on an atomic scale is still in an early stage of development. However, simplified atomic models of surfaces can be made and are useful in explaining the origin, magnitude, and factors influencing surface energy.

A surface is an interface between two phases or two grains of the same phase. All the surface properties of a material depend upon the composition of both of the phases or grains which meet at the interface. Consider the case of the contact of a liquid and its own vapor. In the liquid, the atoms are bound to a number of nearest neighbors, but there is no lattice or long-range pattern to the position of the individual atoms. In the vapor, there is again no long-range order in the position of the atoms and the atoms are considerably farther apart and have greater freedom to move than in the liquid. At the area of transition from vapor to liquid, atoms are bound toward the liquid as they would be in the liquid, and toward the gas only as they would be bound in the gas. Since the atoms, as a result of this condition, have a freedom of movement which is more characteristic of the liquid state, they are considered to be in the liquid, but to possess extra energy because of their partially satisfied bonds toward the vapor. This extra energy manifests itself as an extra attraction between neighboring atoms in the surface (surface tension) and by the potential attraction of foreign elements to the interface (adsorption).

In the liquid-vapor case, at sufficiently high temperature and pressure the transition in bonding from the liquid to the vapor becomes less abrupt until ultimately there is no difference between liquid and vapor states. The excess surface energy goes to zero, and the "critical point" is reached. This liquid-vapor case, in addition to describing the atomic behavior at the critical point, illustrates the general fact that surface tension (and energy) decreases with increasing temperature.

The study of solid-liquid, solid-solid, and solid-vapor interfaces is somewhat complicated by the crystalline nature of many solids and the lack of atomic mobility in solids compared to that in liquids.

## 3.3   UNITS OF SURFACE TENSION AND SURFACE ENERGY

The conventional units employed in describing surfaces are dynes/cm, for surface tension, and ergs/cm$^2$ for surface energy. These units are equivalent to each other since one erg equals one dyne-cm. The fact that force per unit length is equivalent to energy per unit area can be illustrated by the behavior of a soap film on an expandable wire frame (Figure 3.1). If $\gamma$ is the surface tension of the soap-air interface, a force $F = 2\gamma l$ will be required to keep the movable wire stationary (there are two soap-air surfaces resisting $F$). The surface energy is the energy required to increase the surface area. If the wire is moved to the left a dis-

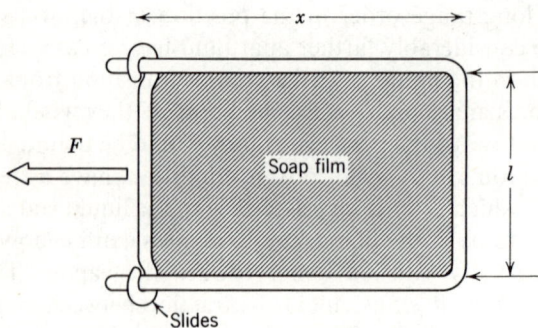

Figure 3.1   Surface tension in a soap film.

tance $dx$, the area $2l\,dx$ will be created at the expense of energy $F\,dx$. The surface energy is then

$$\frac{\text{Energy}}{\text{Area}} = \frac{F\,dx}{2l\,dx} = \frac{2\gamma l\,dx}{2l\,dx} = \gamma \tag{3.1}$$

This relationship illustrates the equality of surface tension and surface energy when the work of creating new surface is solely mechanical. It holds true in most liquids and in solids at temperatures near their melting points. Important divergences from this principle occur in solids at lower temperatures and will be considered in Section 3.5.

### 3.4  MAGNITUDE OF SURFACE ENERGY: METALLIC AND COVALENT MATERIALS

It is difficult to make accurate experimental measurements of surface energy. Where accurate measurements have been achieved, the simplified atomic model of a surface described in Section 3.2 gives a surprisingly good indication of the amount of excess energy associated with a given surface.

Consider the case of a pure crystalline solid in the presence of its own liquid and vapor. There are four surfaces to be taken into account: solid-vapor, solid-liquid, vapor-liquid, and solid-solid (a grain boundary). If the crystal structure of the solid is close packed, the atomic coordination is twelve in the solid, approximately eleven in the liquid, and zero in the vapor. At the solid-liquid surface, a transition in coordination from twelve to eleven occurs; at the solid-vapor surface, a transition from twelve to zero; and at the liquid-vapor surface, from eleven to zero. (The solid-solid surface varies in energy depending upon the orientation of the adjacent grains.) The discontinuity in coordination for the three cases is shown schematically in Figure 3.2. Each surface represents the "breakage" of a certain number of bonds in the denser phase. Similarly, bonds are broken by fusion (solid-liquid), vaporization (liquid-gas), and sublimation (solid-gas). Consequently, we may derive relations between the various surface energies and the energies of fusion, vaporization, and sublimation.

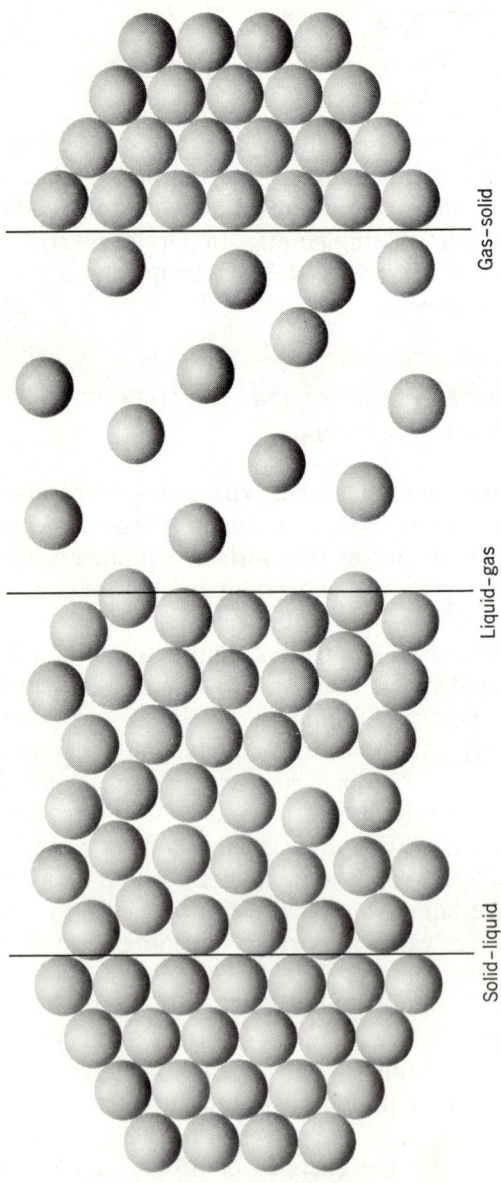

Gas–solid.

Liquid–gas

Solid–liquid

Figure 3.2  Schematic diagram of coordination change of atoms at solid-liquid, liquid-gas, and gas-solid surfaces.

The solid-vapor surface energy $\gamma_{SV}$ is approximately equal to the energy (per unit area) of the unsatisfied bonds produced by the transfer of an atom from the interior of the material to its surface. If the interface is a close-packed plane of the solid, there will be three half bonds per atom toward the vapor. Sublimation of each atom (from the surface) requires the breaking of nine bonds between the atom and the interior of the solid. Thus the sublimation energy $\Delta H_s$ represents the energy of formation of eighteen half bonds. The ratio of $\gamma_{SV}$ to $\Delta H_s$ should be $3:18$, or about 0.166. Table 3.1 shows this ratio for several metallic elements for which sufficient data are available. The values range from 0.129 to 0.241 and thus are in reasonable agreement with the predicted value of 0.166.

The energy of the liquid-vapor surface $\gamma_{LV}$ may be derived by recognizing that liquid coordination is 11/12 of solid coordination. We would therefore expect that $\gamma_{LV}$ would be about 11/12 of $\gamma_{SV}$. Table 3.1 shows values from 0.75 to 0.96, which are in agreement with the derived ratio. Similarly, solid-liquid surfaces represent a transition from a coordination of twelve to a coordination of eleven, a change of one compared with a change of twelve for the solid-vapor case. The solid-liquid surface energy, therefore, might be expected to be one order of magnitude less than the solid-vapor surface energy and to be only a fraction of the heat of fusion

Table 3.1   Surface Energies and Heats of Transformation for Several Metals

| | SILVER | GOLD | NICKEL | COPPER |
|---|---|---|---|---|
| Atomic diameter, Å | 2.880 | 2.878 | 2.486 | 2.551 |
| Atomic area on (111) plane, cm² | $7.18 \times 10^{-16}$ | $7.18 \times 10^{-16}$ | $5.35 \times 10^{-16}$ | $5.65 \times 10^{-16}$ |
| $\Delta H_s$ kcal/mole | 82 | 60 | 114 | 73.3 |
| $\Delta H_f$ kcal/mole | 2.27 | 3.11 | 4.2 | 2.67 |
| $\gamma_{SV}$ ergs/cm² | 1200 | 1400 | 1900 | 1700 |
| $\gamma_{SL}$ ergs/cm² | 126 | 132 | 255 | 177 |
| $\gamma_{VL}$ ergs/cm² | 920 | — | 1825 | 1270 |
| $\gamma_{SV}/\Delta H_s$ | 0.151 | 0.241 | 0.129 | 0.188 |
| $\gamma_{LV}/\gamma_{SV}$ | 0.766 | — | 0.96 | 0.75 |
| $\gamma_{SL}/\gamma_{SV}$ | 0.105 | 0.044 | 0.134 | 0.104 |
| $\gamma_{SL}/\Delta H_f$ | 0.505 | 0.435 | 0.196 | 0.540 |

$\Delta H_f$. The energy of the solid-solid or grain boundary surface depends upon the angles of the adjacent grains relative to one another. Experiments indicate that it is about one third of the solid-vapor surface energy in a given material.

Table 3.1 lists the surface energies of several metals. The values for other metals range from 500 ergs/cm². Oxides frequently have lower surface energies; that of water at room temperature is about 70 ergs/cm².

It is important to realize that the *broken-bond* assumption applies only to materials which are bonded by *short-range* forces. This restriction is necessary for the complete bonding energies to be realized after removal of an atom to or from a surface. Consequently, the considerations of this section should apply well to materials bonded metallically or by covalent forces.

## 3.5  MAGNITUDE OF SURFACE ENERGY: IONIC MATERIALS

The atoms in ionic solids are bonded by coulomb forces which are essentially long range in nature. We cannot assume, as we did in section 3.4, that an atom removed from a solid surface totally escapes the influence of the surface. Consider, however, the creation of two surfaces by mechanically separating an ionic crystal into halves. The surface energy thus created must be numerically equal to the work done, which is

$$\gamma_{SV} = \int_0^\infty \sigma \, dx \qquad (3.2)$$

where $\sigma$ is the stress (force per unit area) used, and $x$ is the separation of the two halves. The potential energy of the ionic lattice is periodic, repeating itself every interatomic distance $2r_0$. The simplest periodic function is a sine or cosine. For a first, crude approximation we assume that separating the two halves of the crystal means removal of the material from a sinusoidal "potential well." After the first potential well, the potential energy will further be assumed to be zero. The force necessary to maintain separation between the two halves will also be sinusoidal over the first cycle, rising from zero at zero separation to some maximum, then falling off to zero again at separation $r_0$. Therefore

$$0 \leq x \leq r_0 \qquad \sigma = \sigma_0 \sin\left(\frac{2\pi x}{2r_0}\right) \qquad (3.3)$$

$$r_0 < x \qquad \sigma \sim 0$$

In the limit of zero separation the laws of elasticity apply. The stress must be proportional to the separation (*Hooke's Law*). In particular, the elastic constant $E$, called *Young's Modulus*, is known:

$$E \equiv \left[\frac{\partial \sigma}{\partial \epsilon}\right]_{\epsilon=0} = \left[\frac{d\sigma}{d(x/r_0)}\right]_{x=0}$$

Also, from equation 3.3,

$$\left[\frac{d\sigma}{d(x/r_0)}\right]_{x=0} = \left[\pi\sigma_0 \cos\left(\frac{\pi x}{r_0}\right)\right]_{x=0} = \pi\sigma_0$$

So that

$$\sigma_0 = \frac{E}{\pi}, \qquad \sigma = \frac{E}{\pi} \sin\left(\frac{\pi x}{r_0}\right)$$

and therefore

$$\gamma_{SV} = \frac{E}{\pi} \int_0^{r_0} \sin\left(\frac{\pi x}{r_0}\right) dx = \frac{E r_0}{\pi^2} \qquad (3.4)$$

Since $r_0$ is also known, from x-ray and other data, $\gamma_{SV}$ may be calculated. Table 3.2 compares calculations of $\gamma$ according to equation 3.4 with the results of direct measurements of $\gamma$, which will be discussed in the next section.

*Table 3.2    Calculated and Experimental Surface Energy Values for Various Ionic Materials*

| MATERIAL | $\gamma_{SV}$(EXP.) | $\gamma_{SV}$(CALC.) |
|---|---|---|
| NaCl | 300 ergs/cm$^2$ | 310 ergs/cm$^2$ |
| MgO | 1200 | 1300 |
| LiF | 340 | 370 |
| CaF$_2$ | 450 | 540 |
| BaF$_2$ | 280 | 350 |
| CaCO$_3$ | 230 | 380 |

## 3.6 MEASUREMENT OF SURFACE ENERGY

The most direct method of measurement is to determine the heat of solution (or some reaction) of very fine particles of known size. As a result of the added surface energy of the particles, the heat of solution will differ from that of the bulk material. This difference, together with the surface area of the particles, is sufficient for the direct calculation of the surface energy.

If liquid surfaces are involved, "wetting" of a solid surface may be used. A liquid spreads along a solid surface rather than forming a spherical drop when

$$\gamma_{SL} + \gamma_{LV} < \gamma_{SV} \tag{3.5}$$

that is, when the net free energy is lowered by replacing an S-V surface by an S-L and an L-V surface together. On the other hand, no wetting at all will occur if

$$\gamma_{SV} + \gamma_{LV} < \gamma_{SL} \tag{3.6}$$

Wetting and nonwetting are shown in Figure 3.3 together with the case of partial wetting, which allows quantitative calculation of the surface energies. In the plane of the solid surface force equilibrium must exist between the three surface tensions because the

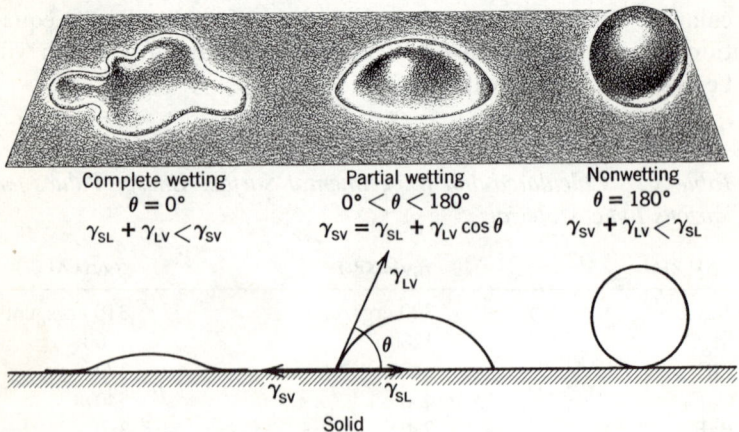

| Complete wetting | Partial wetting | Nonwetting |
|---|---|---|
| $\theta = 0°$ | $0° < \theta < 180°$ | $\theta = 180°$ |
| $\gamma_{SL} + \gamma_{LV} < \gamma_{SV}$ | $\gamma_{SV} = \gamma_{SL} + \gamma_{LV}\cos\theta$ | $\gamma_{SV} + \gamma_{LV} < \gamma_{SL}$ |

Figure 3.3   Wetting of a substrate by a liquid.

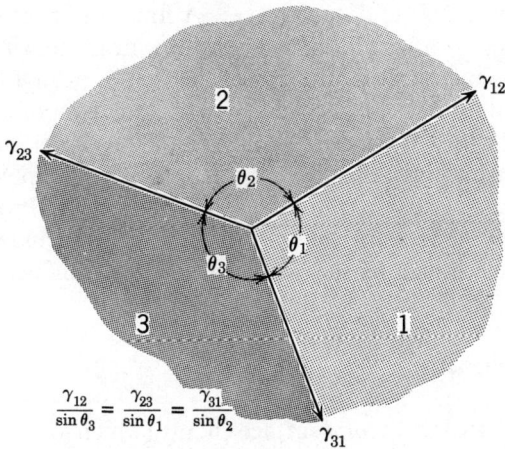

$$\frac{\gamma_{12}}{\sin \theta_3} = \frac{\gamma_{23}}{\sin \theta_1} = \frac{\gamma_{31}}{\sin \theta_2}$$

Figure 3.4    Force equilibrium between phases.

liquid droplet is free to move until force equilibrium is established.  Hence,

$$\gamma_{SV} = \gamma_{SL} + \gamma_{LV} \cos \theta \qquad (3.7)$$

Force equilibrium may also be used when all three phases are mobile, so that mechanical equilibrium is established in three dimensions.  Figure 3.4 describes the situation that exists in any plane which cuts the junction of three such phases.  At high temperatures atomic mobility in solids is high enough to justify the use of the force equilibrium for solid materials:

$$\frac{\gamma_{12}}{\sin \theta_3} = \frac{\gamma_{23}}{\sin \theta_1} = \frac{\gamma_{31}}{\sin \theta_2} \qquad (3.8)$$

The force-balance methods of equations 3.7 and 3.8 yield only ratios, not absolute values of $\gamma$.  If one absolute value is known, complete solution is possible for equation 3.8.  Similarly, two of the three surface energies in equation 3.7 must be known.  The reaction or solution of fine powders yields absolute values for $\gamma$. However, the powders must be strain-free and completely free of surface contamination.  It is almost impossible to meet the two conditions simultaneously.  However, there is another technique which gives absolute data for $\gamma_{SV}$.  A fine, single-crystal wire is used to support a weight, at temperatures near the wire's melting point.  The mechanical strength of the wire is negligible, and the weight is supported almost wholly by the solid-vapor surface tension of the wire.  If the wire is polycrystalline, the grain boundary surface tension must also be considered.  The complete solution to the wire-suspension problem is carried out in Problems 3.7, 3.8, and 3.9.  The solid-vapor surface energies included in Table 3.1 were determined by the wire-suspension technique.


## 3.7  ADSORPTION

There are two ways *total* surface energy (which must be included in the free energy) may be reduced: reduction of surface area and reduction of surface energy per unit area ($\gamma$).  Both processes will occur.  The phases will adjust their shapes until, at equilibrium, surface area is minimized.  The composition of each surface will also be adjusted so that $\gamma$ is minimized.  Thus the surface will, in general, differ in composition from the bulk.  We may calculate the extent of the difference by using free energy.  First, however, the qualitative statement: those impurities or components which reduce $\gamma$ will segregate to the surface; those which increase $\gamma$ will segregate away from the surface.  Figure 3.5 is a typical plot of $\gamma$ versus concentration for a completely miscible binary system. The surface tension is constant, roughly $\gamma_B$, from 50% B to pure B

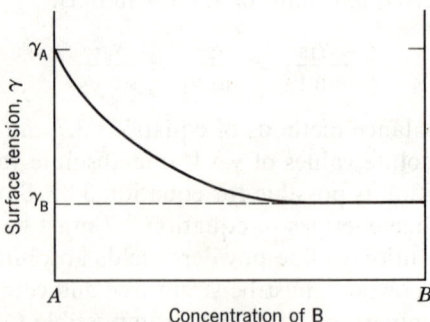

Figure 3.5  Surface tension versus composition in a completely miscible binary system.

because the surfaces of the solid are essentially pure B.  When A appears in the surface, the surface tension is raised.

It is also important to note that only a tiny fraction of the total mass of any material is associated with the surfaces.  Similarly, tiny amounts of impurities are sufficient to saturate the surfaces.  On a grosser scale, 10% oxygen in the atmosphere reduces $\gamma_{SV}$ for silver from 1200 to 400 ergs/cm$^2$.  On a finer scale, it has been found that really clean metallic surfaces, free of adsorbed gases, may be maintained only in an extremely good vacuum; usually, the total pressure must be below $10^{-10}$ mm. of mercury ($10^{-13}$ atmosphere)!  Because so very little material is required, the grain boundaries of solids usually are saturated thoroughly with impurities.  Also, for the same reason, the composition of any phase boundary may change with negligible effect on the compositions of the bulk phases.  We may therefore write, for any surface,

$$dF^{(s)} = \gamma \, ds + \sum_{i=1}^{i=c} \frac{\partial F^{(s)}}{\partial n_i^{(s)}} \, dn_i^{(s)} \tag{3.9}$$

in the same manner as we wrote equation 2.12.  The term $\gamma \, ds$ is the change in energy due to the change in surface area $ds$.  The summation index $i$ refers to summation over all the components.

The symbols $F^{(s)}$ and $n_i^{(s)}$ refer to the Gibbs Free Energy of the surface and the number of moles of component $i$ in the surface, respectively.  The term $\partial F^{(s)}/\partial n_i^{(s)}$ is measurable and is usually referred to by the symbol $\mu_i^{(s)}$.  In Problem 3.10, we show that

$$\frac{n_i}{s} = -\frac{d\gamma}{d\mu_i} \tag{3.10}$$

Thus, from measurements of surface energy and free energy, the segregation of any component to or from a surface may be predicted.  Usually $\gamma$ is expressed as a function of concentration and is substituted into equation 3.10, which is called the *Gibbs Adsorption Isotherm*.

## 3.8   FURTHER REMARKS ON SOLID SURFACES

The surface energy $\gamma_{SV}$ depends on crystallographic orientation.  A single crystal held at elevated temperatures will always assume a shape bounded by crystallographic planes of minimum surface energy. This is not true of liquids, whose surface energy is isotropic. Using the *broken-bond* technique of Section 3.4, we can show that the close-packed planes have the lowest energy per unit area. (See Problem 3.3.) Thus, the calculations of Section 3.4 and Table 3.1 use only the {111} planes of the four metals, which are all face-centered cubic. Similarly, in ionic crystals the balance of coulomb forces leads to anisotropy of $\gamma_{SV}$, and chemical reactivity is anisotropic for crystalline solids. Figure 3.6 is a photograph which demonstrates such behavior. The triangular "etch pits" are bounded by the {110} planes because the etching process is least rapid on that particular family of crystallographically identical planes.

Another consequence of anisotropy in solids is that the surface energy and the surface tension are no longer exactly the same. In Section 3.3, we justified the assumption that the entire force exerted by the soap film was in the plane of the film. In solids, however, the assumption does not apply. The force on the surface need not be in the plane of the surface and furthermore may

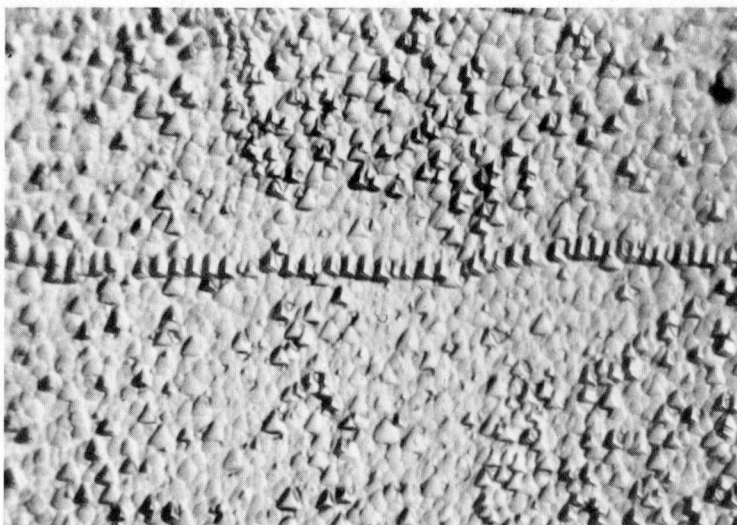

Figure 3.6   Etch pits developed on a {111} surface of a niobium single crystal, due to anisotropy of chemical reactivity. Courtesy of C. S. Tedman, Jr.   1000X.

depend on the crystallographic orientation of the surface. The surface tension then becomes a complicated quantity; mathematically such a quantity is called a tensor. However, experimental data on surfaces, in most cases, are not presently refined enough to require such complications.


## DEFINITIONS

*Adsorption.* The segregation of components to or from a surface.
*Surface.* The boundary between two phases.
*Surface Energy* ($\gamma$). The excess energy per unit area associated with a surface; the energy needed to increase surface area by one unit of area.
*Surface Tension.* The tendency to minimize total surface energy by minimizing surface area.
*Wetting.* The spontaneous spreading of one phase over the surface of another.


## BIBLIOGRAPHY

Gilman, J. J., *Journal of Applied Physics,* vol. 31, p. 2208 (1960). The calculation of the surface energy of ionic crystals.

Gomer, R., and C. S. Smith, (Editors), *Structure and Properties of Solid Surfaces,* University of Chicago Press, 1953. This volume, containing a series of papers on research and the state of knowledge of surfaces, requires some familiarity with the field.

Kingery, W. D., *Property Measurement at High Temperatures,* Wiley, New York, 1959. A good review of the experimental techniques for surface energy measurement is included.

Swalin, R. A., *Thermodynamics of Solids,* Wiley, New York, 1962. Chapter 12 is a thorough account of interface thermodynamics on the undergraduate level.

Udin, H., A. J. Shaler, and J. Wulff, *Journal of Metals,* vol. 1, p. 1936 (1949). The measurement of surface energy by the hanging wire method.


## Problems

3.1 What are the units for surface tension and surface energy? In what way and under what conditions are they equal?

3.2 What is the origin of surface energy?

3.3 Show that the close-packed plane in a FCC crystal has the minimum surface energy. (Hint: the calculations in Table 3.1 were made for the close-packed {111} plane.  Find the number of unsatisfied bonds per unit area on this plane and on several other low-index planes.)

3.4    The atomic diameter of silver is 2.88 A; its crystal structure is FCC.  Estimate, using the data in Table 3.1, the surface energy of the {100} plane.

3.5    What is the usual effect of impurities on the surface energy of a material and why?

3.6    The case of partial wetting of a solid by a liquid is illustrated in Figure 3.3.  If a droplet such as that shown in this figure is placed on an inclined solid surface so that it moves downward, it has been found that the values of $\gamma$ at the advancing interface is different from $\gamma$ at the receding interface.  Suggest an explanation of this fact.

3.7    Show that the force necessary to balance the surface tension of a fine wire at elevated temperature is given by

$$f = mg = \pi r \gamma - \pi r^2 \left(\frac{n}{l}\right)\gamma_{GB}$$

where $r$ is the radius of the wire, $n/l$ is the number of grains per unit wire length, and $\gamma_{GB}$ is the surface energy of grain boundaries.

3.8    A nickel wire 0.005 inches in diameter is suspended in its own vapor at 1300°C.  It is found that a weight of 0.0366 gm is required to balance the tendency of the wire to shrink, and that there are thirty grains per centimeter of wire length.  Calculate the surface energy (assuming that $\gamma_{GB} = \frac{1}{3}\gamma_{SV}$).

3.9    What error would result in the calculation of surface energy if the wire of Problem 3.8 were assumed to be a single crystal?

3.10    Derive equation 3.10 from equation 3.9.  Integrate equation 3.9 to obtain

$$F^{(s)} = \gamma S + \sum_{i=1}^{c} \mu_i^{(s)} n_i^{(s)}$$

Explain the path of integration, that is, explain why $\gamma$ and $\mu_i^{(s)}$ may be taken as constants.  Differentiate the above relation, and subtract equation 3.9.

3.11    (a) Sketch the apparatus needed to measure the wetting of a solid metal by a liquid metal above 1000°C.

(b) What surface energy relationship determines the wetting behavior and the spreading tendency of one metal on another?

3.12    (a) Define sintering of metal powders.

(b) What is the driving force for densification during sintering?

(c) Describe an evaporation-condensation sintering process using simple sketches.

3.13   (a) Sketch the area of contact between two real surfaces.

(b) How is the "roughness" of a really fine surface measured and in what units?

(c) What is the size of the real area of contact?

(d) What is the magnitude of the friction force if we try to slide the two surfaces over one another?

3.14   Write a brief critical essay on the "adhesional" theory of friction and include remarks about the nature of adhesive wear.

3.15   (a) Define chemical adsorption (chemisorption).

(b) Distinguish (a) from physical or van der Waals adsorption.

(c) How can it be shown that chemisorbed films of oxygen are more tightly bonded than oxide films?

(d) Why should a chemisorbed film be responsible for the passivity of stainless steel rather than a true oxide film?

CHAPTER FOUR

# Rates of Reactions

SUMMARY

Although a reaction may be thermodynamically favored, the attainment of equilibrium may be impeded by various physical factors. The *Arrhenius Equation,* which originated empirically, leads to the concept of an activation energy, an *energy barrier* which must be surmounted in order to proceed to equilibrium. If thermal energy must be supplied for activation, the probability of acquiring this energy may be computed by *Maxwell-Boltzmann* statistics, and the Arrhenius Equation then follows. The effect of temperature and catalysis on the rate of reactions such as diffusion, nucleation, growth, and oxidation may be interpreted and predicted in this way.

## 4.1 INTRODUCTION

A large class of transformations in materials, although thermodynamically possible, occur very slowly. Often materials may remain for long periods of time in nonequilibrium states. Wood provides an example of such a *metastable* form; various metals which have higher free energies than their respective oxides, but which exist under suitable conditions for long periods of time before oxidizing, provide another. This apparent lack of spontaneous reaction is in fact just a very slow reaction *rate*. The rates of reactions are controlled by the existence and nature of any barriers retarding an approach to equilibrium. Since all processes are affected to a greater or lesser degree by the presence of such barriers, a general way of studying them has been evolved. The study of rates of reaction or transformation is known as *kinetics* or *rate process theory*. Although thermodynamic considerations establish the *possibility* of a reaction, kinetic principles govern the *rate* of that reaction.   In this chapter, the general characteristics of rate processes will be considered.   Later chapters will include a more detailed, individual treatment of a number of specific rate processes: diffusion, nucleation and growth in phase transformations, sintering, oxidation, and corrosion.

## 4.2   THE ARRHENIUS EQUATION

The Swedish chemist Arrhenius (1859–1927) observed that the increase in rate of chemical reactions with increased reaction temperature could usually be expressed by

$$\text{rate} = \text{constant} \times e^{-Q/RT} \tag{4.1}$$

in which $Q$ is the *activation energy* in units of calories per mole, $R$ is the gas constant (1.98 cal/mole °K), $T$ is the absolute temperature °K, and the rate constant is independent of temperature. A large number of reactions and transformations, both nonchemical and chemical, can be described by equation 4.1, the *Arrhenius Equation,* which is commonly used for analyzing experimental rate data.

A process of any kind—diffusion, change of composition, and so on—may be analyzed kinetically by measuring the rate at which it occurs.   The logarithm of the rate can then be plotted as a function of the reciprocal of the absolute temperature at which the rate prevailed.   If this experimental plot proves to be linear, the resulting graph resembles Figure 4.1.   The equation of the line in Figure 4.1 is

$$\log_{10} \text{rate} = \log_{10} (\text{constant}) - \frac{Q}{2.303R}\left(\frac{1}{T}\right) \tag{4.2}$$

This is exactly the relation obtained by taking logarithms of the Arrhenius Equation.   Here 2.303 is the factor of conversion between common logarithms and natural logarithms.   The activation energy may be computed from the data in Figure 4.1 by recognizing that the slope of the line is:

$$\text{slope} = -\frac{Q}{2.303R} \tag{4.3}$$

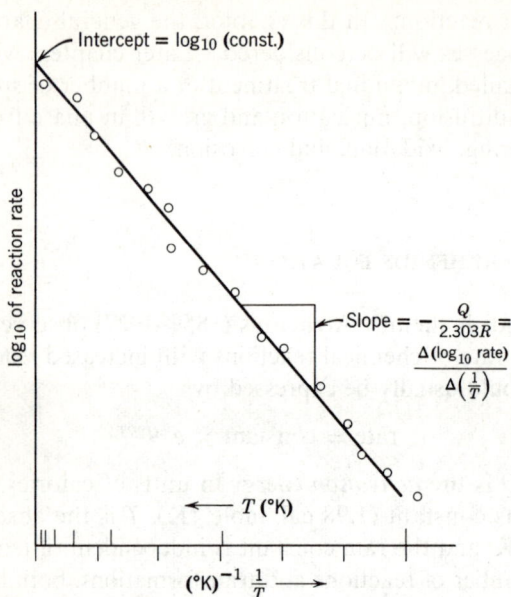

Figure 4.1   Typical Arrhenius plot of experimental rate data.

The rate coefficient may be evaluated by extrapolating the rate line in Figure 4.1 to the point at which $1/T$ equals zero. This type of calculation is illustrated in Problem 4.1.

The rate of a reaction involving a succession of steps is controlled by the slowest step. The values of $Q$ and the rate coefficient determined by experimental measurements are useful, when combined with other types of observations, for identifying such *rate-controlling processes*. Of course, several alternative processes may have the same activation energies, and then other experimental techniques must be used to distinguish between them.

## 4.3   ACTIVATION ENERGY

In equation 4.1 the dependence of reaction rate upon temperature was specified by the quantity $Q$ having typical units cal/mole. The exponential dependence of temperature resembles that of the

*Maxwell-Boltzmann Distribution,* which specifies the energy distribution of molecules in gases. The Boltzmann relation expresses the probability of finding a molecule at an energy $\Delta E$ greater than the average energy at a particular temperature $T$:

$$\text{probability} \propto e^{-E/kT} \tag{4.4}$$

In this form $\Delta E$ is an energy in ergs per molecule and $k$ is the Boltzmann or molecular gas constant in ergs/molecule/°K.

The similarity in form between the experimentally derived Arrhenius Equation and the Boltzmann relation (equation 4.4) suggests an important property of the rate of a reaction or transformation. Apparently the reaction rate depends on the number of reacting species that have an amount of energy $\Delta E^*$ greater than the average energy $E_r$ of the reactants. The energies of reactants and products in such a case are illustrated schematically in Figure 4.2. In this figure, the curve represents the energy of a single reacting species as it progresses along a *reaction coordinate* from the unreacted condition on the left to the reacted condition on the right. For instance, the reacting species might be an atom diffusing through a crystal lattice. Each minimum in Figure 4.2 would then be an equilibrium site; the maximum would be the position between the sites, where neighboring atoms must be pushed aside in order to pass from one site to another. The reaction coordinate would be the physical distance traveled by the atom. Another example is the dissociation of a metastable molecule where the coordinate would be the interatomic distance. The atoms need a certain amount of additional energy, usually because of some short-range attractive forces, before they can become separated. The reaction coordinate need not be an actual physical distance but instead some measure of the extent of reaction. The unit of any reaction or transformation may be visualized on the reaction coordinate and the energy of the system expressed as a function of position on the reaction coordinate.

The rate at which the reacting species of Figure 4.2 passes over the activation barrier is determined by the height of that barrier, $\Delta E^*$ from left to right, and the source of energy available to the reacting species. If thermal energy alone is available, the probability of acquiring $\Delta E^*$, and thereby surmounting the barrier, is given by equation 4.4.

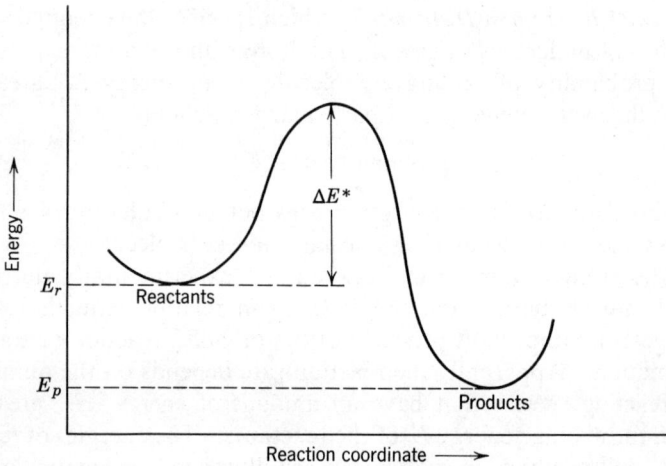

Figure 4.2   Energy of reacting species as it proceeds from the unreacted to the reacted state.

A qualitative mechanical model assists in visualizing the energy changes involved.   In Figure 4.3 a rectangular box is oriented in three positions of different potential energies.   The box when in position (*a*) possesses a higher potential energy than when in posi-

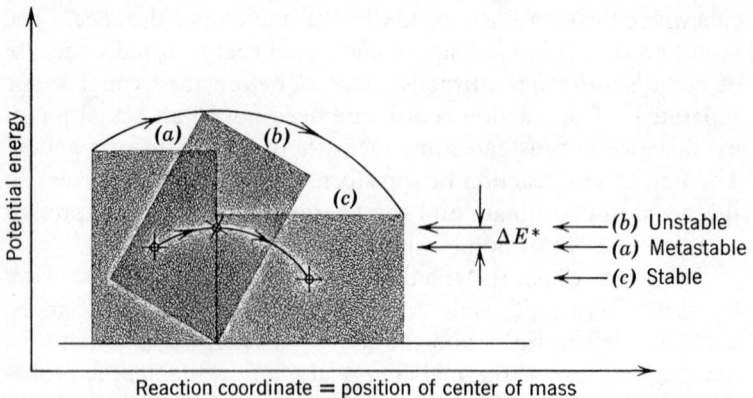

Figure 4.3   A mechanical analog to metastability and activation.

tion (c). If nothing pushes the box, it remains in position (a) indefinitely. Since the potential energy of position (c) is less than that of position (a), position (a) is *metastable* with respect to position (c). If the potential energy of the box is increased by pushing it into position (b), its position will be *unstable*. The excess potential energy of position (b) over that of position (a) corresponds to the activation energy of the reaction from (a) to (c).

## 4.4 THERMODYNAMICS OF KINETICS

In Chapter 1 it was demonstrated that free energy $F$ is minimized at equilibrium. Since free energy is computed by using both the enthalpy and entropy terms, it is clear that both of these qualities must be recognized in studying any actual process. Similarly, in the box example of Figure 4.3 only the contribution of the potential energy to enthalpy was considered. Not taken into account was the fact that all the atoms in the block would have to be thermally excited to move in the same direction, at the same time, before the block would turn over spontaneously. The probability of this simultaneous movement is very small and, therefore, the block does not turn over from thermal excitation alone. This low probability of reaction is due to the large entropy change that would have to accompany the process. For this reason the probability expression (equation 4.4) should be written in terms of the free energy of activation per mole $\Delta F^*$ rather than internal energy, as follows:

$$\text{probability} \propto e^{-\Delta F^*/RT} \tag{4.5}$$

Since $\Delta F^* = \Delta H^* - T \Delta S^*$

$$\text{probability} \propto \{e^{\Delta S^*/R}\}\{e^{-\Delta H^*/RT}\} \tag{4.6}$$

This more rigorous expression for the probability of occurrence of a reaction has several important effects on the study of reaction rates discussed in Section 4.3. The energy coordinate in Figure 4.3 is now seen to be, specifically, a free energy coordinate. This means that the concept of enthalpy alone is not sufficient to describe the rate of a reaction, but that changes of entropy are also significant and particularly when large numbers of atoms must react simultaneously. This becomes important in concepts of nucleation to be developed in Chapter 6.

When the free energy barrier is used to describe the rate of a reaction, as in equation 4.5, the result is completely consistent with the free energy criterion for equilibrium considered in Chapter 1. In Figure 4.4 the progress of a reaction is sketched. Free energy is used as the measure of stability. In this figure the reaction from ① to ② is thermodynamically favored since $\Delta F_{12}$ is negative. The rate of ① and ② would be governed by the "barrier" $\Delta F_{12}*$. The reverse reaction, ② to ①, is not favored thermodynamically since $\Delta F_{21}$ is positive; however, in principle it would still proceed at a rate governed by $\Delta F_{21}*$. The net reaction rate is the difference between competing rates ① to ② and ② to ①:

$$\text{net rate} \propto n_1 e^{-\Delta F_{12}*/RT} - n_2 e^{-\Delta F_{21}*/RT} \tag{4.7}$$

where $n_1$ is the number of units (e.g., atoms, molecules, nuclei) in state ①, and similarly $n_2$ is the population of state ②. When $\Delta F_{12}$ is negative, the reaction rate from ① to ② exceeds that of ② to ①. When $\Delta F_{12}$ is positive, the reverse reaction ② to ① occurs more rapidly, and the role of reactants and products is reversed. When the net rate of the reaction is zero, thermodynamic equilibrium prevails; the competing rates are equal. This concept of

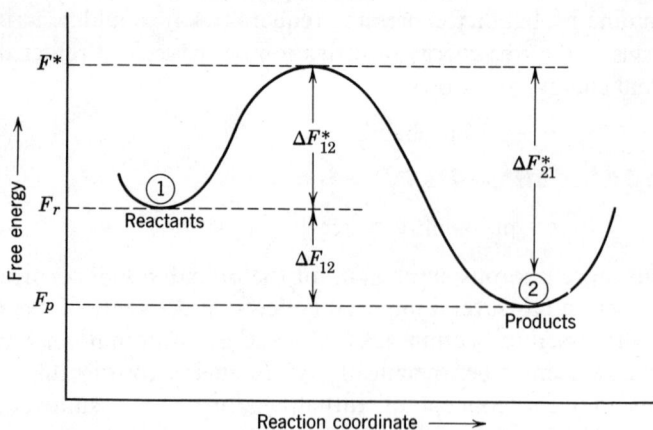

Figure 4.4   Free energy path of a thermodynamically favored reaction.

*dynamic equilibrium* is essential to the discussion of diffusion in Chapter 5.

Setting equation 4.7 equal to zero immediately gives

$$\frac{n_2}{n_1} = e^{(\Delta F_{21}{}^* - \Delta F_{12}{}^*)/RT} = e^{-\Delta F_{12}/RT} \tag{4.8}$$

This is exactly the thermodynamic equation for equilibrium in a system described by Figure 4.4. In chemical systems $n_2/n_1$ is called the *equilibrium constant*. Equation 4.8 is perfectly general. In addition to chemical equilibrium, it describes electron densities in the states of a maser, the magnetic moment distribution in atomic and molecular aggregates, semiconductor and insulator leakage and breakdown, and any other situation where thermal energy is a source of activation energy.

The conclusions of Sections 4.2 and 4.3 must now be modified by the thermodynamic concepts which were included in equations 4.5, 4.6, and 4.7. The Arrhenius Equation describes a reaction at some distance from equilibrium only. An example of this is $\Delta F_{21}{}^* > \Delta F_{12}{}^*$ in Figure 4.4 and equation 4.7. The activation energy, $Q$ of the Arrhenius Equation, is really the activation enthalpy, as in equation 4.6. The activation enthalpy describes only the temperature dependence of a reaction rate and does not include entropy changes. These are, therefore, to be found in the pre-exponential constant of the Arrhenius Equation when experimental data are evaluated. It is the effect of entropy which makes the complete calculation of rates from first principles difficult, at the present state of knowledge.

## 4.5   MODIFICATION OF REACTION RATES

Although it is not possible to evaluate *all* of the factors affecting the rate of a reaction, it *is* possible to evaluate some of them. Once the nature of their influence is known, such factors may be varied to change a reaction rate to practical advantage. In subsequent chapters several such possible variables will be considered; there are two variables of such a general nature, however, that they are considered here.

*Temperature* is the most obviously controllable variable in equation 4.6.   A popular statement concerning chemical and physical reactions is that the rate of reaction doubles for every ten-degree rise in temperature.   This rough estimate is merely another way of saying that the apparent activation enthalpies of many practical processes lie between 10,000 and 100,000 cal/mole. Calculations such as those in Problem 4.4 demonstrate this.

*Activation free energy* is the other variable in equations 4.5 or 4.6 which may be modified in order to change the reaction rate. A lower $\Delta F^*$ value produces a higher reaction rate (provided that other variables are held constant).   This leads to the generalization that of a number of alternative paths for reaction, the one having the lowest free energy barrier will occur most rapidly.   In some chemical reactions the presence of a catalyst, which does not itself enter into the chemical reaction is essential to a rapid reaction rate.   The catalyst offers an alternative reaction path which has a lower activation enthalpy barrier, as shown in Figure 4.5. Usually, it is the surface of the catalyst which performs this function.   Under certain circumstances the action can be "poisoned"

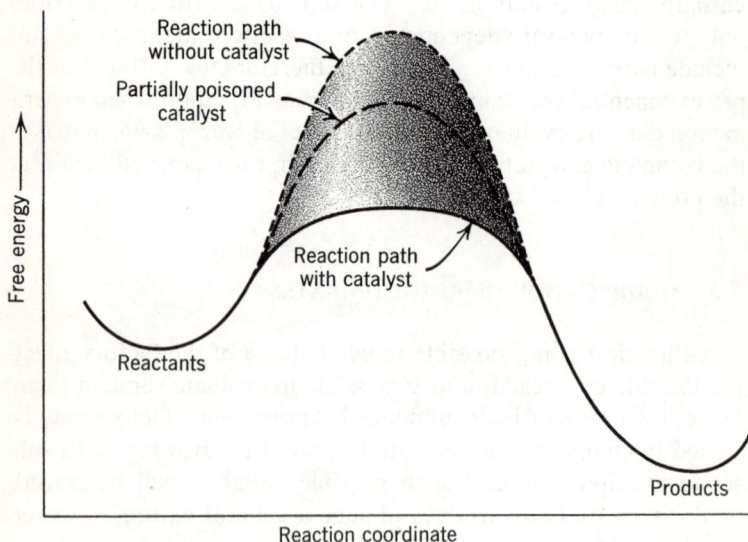

Figure 4.5   Influence of catalysis on the free energy path of a reaction.

by impurity atoms which shield the surface from the reactants. When this occurs, the advantage of the catalyst is diminished, the free energy path of the reaction assumes a progressively higher position in Figure 4.5, and the rate of reaction slows down. If, however, the "poison" were removed from the catalyst, or if there were none there to begin with, the catalyst would not be consumed during the reaction and would continue to be useful.

A catalyst simply offers a suitable surface as a low activation energy site for a chemical reaction. Other examples of the influence of surfaces on reaction rates will be considered in Chapter 6.

## DEFINITIONS

*Activation Energy.*  The height of the energy barrier to a reaction which must be surmounted by thermal excitation.

*Activation Free Energy.*  The free energy barrier to a reaction, including the changes in entropy and enthalpy.

*Arrhenius Equation.*  An empirically based equation describing the rate of a reaction as a function of temperature and any energy barriers to the reaction.

*Catalyst.*  A substance which does not enter into a reaction but whose presence lowers the activation energy for that reaction.

*Kinetics, Rate Process Theory.*  The study of reaction rates and the factors influencing them.

*Rate-Controlling Process.*  The individual step in a reaction involving a series of steps which occurs at the lowest rate and which, therefore, establishes the rate of the reaction as a whole.

## BIBLIOGRAPHY

Daniels, F., and R. A. Alberty, *Physical Chemistry,* Wiley, New York, 1961.

Glasstone, S., K. J. Laidler, and H. Eyring, *Theory of Rate Processes,* McGraw-Hill, New York, 1941.

Hinshelwood, C. M., *Kinetics of Chemical Change,* Oxford University Press, London, 1940.

## Problems

4.1  The time which elapses before any evidence of the reaction A → B is obtained has been investigated as a function of temperature. The data are:

| TIME | TEMPERATURE |
|---|---|
| 77 minutes, 50 seconds | 327°C |
| 13.8 seconds | 427°C |
| 0.316 second | 527°C |
| 1 millisecond | 727°C |

What is the activation energy for this reaction?

4.2    The time in seconds necessary for severely deformed copper to soften by 50% is given by

$$t = 10^{-12}e^{30,000/RT}$$

where $R$ is in calories per mole per degree Kelvin and $T$ is in degrees Kelvin. At 1000°C, how long will the copper take to soften by 50%? If the copper is left at room temperature, how much time is needed? Suppose the copper is being hot-worked, and 50% softening must occur in one second in order to keep working stresses down. What should be the minimum processing temperature?

4.3    The following tautology is notorious: "The most stable state is invariably the *last* one to occur." Is this statement totally meaningless? Discuss.

4.4    Suppose the absolute rate of a certain process cannot be measured; however, the *ratio* of the rates, $r_2/r_1$, at the two temperatures $T_2$ and $T_1$ can be measured. Find the activation energy of this process in terms of $T_2$, $T_1$, and $r_2/r_1$.

4.5    Mr. S is very particular about his roast beef. Mrs. S finds that she can roast the meat at higher temperatures for shorter times or at lower temperatures for longer times. The combination of time and temperature necessary to prepare the roast exactly to Mr. S's taste is

$$T \ln t = \text{constant}$$

Explain Mrs. S's observation. What assumptions must be made?

4.6    The following reaction rate data have been obtained:

| $T$ | 227°C | 327°C | 377°C | 427°C | 527°C |
|---|---|---|---|---|---|
| Rate | 1.00 | 6.61 | 14.0 | 25.7 | 128 |

Analyze these data. Explain what is occurring.

4.7    The free energy of activation $\Delta F^*$, for a particular phenomenon, has been found to be

$$\Delta F^* = 23,000 - 8.1T - 1.5 \times 10^{-4}T^2$$

Use the Gibbs-Helmholtz equation (Problem 1.23) to find $\Delta H^*$. Could reaction rates for this system be analyzed in the classical manner (i.e., by the Arrhenius Equation)? Why or why not? If there appears to be a better method of analysis, describe it.

4.8   From equation 4.5 and the Gibbs-Helmholtz equation (Problem 1.23) prove that

$$\frac{\partial \ln r}{\partial T} = \frac{\Delta H^*}{RT^2}$$

($r$ = rate of process)

4.9   Plastic flow and fracture of a large class of materials are thermally activated; however, activation of such processes is aided by the applied stress. For instance, when the applied stress causes yielding, it also does work. This work is subtracted from the activation energy for the unstressed case because it is supplied by mechanical means and not by thermal energy and, therefore, should not be included in the expression for temperature dependence. In addition, the activation energy (or enthalpy) is usually a function of temperature; the atomic mechanisms of flow and fracture are usually very temperature-sensitive. To investigate these phenomena, the stress level ($\sigma$) necessary to produce a given constant strain is expressed as a function of strain rate ($\dot{\epsilon}$) and temperature ($T$). The *reaction rate* equation then becomes

$$\dot{\epsilon} = A e^{[G\Delta H^*(T,\sigma)]/RT} = A e^{(-\Delta S^*)/R} e^{[-\Delta F^*(T,\sigma)]/RT}$$

where $A$ is independent of $\sigma$, $T$, and $\dot{\epsilon}$. The activation enthalpy is generally assumed to depend linearly on stress.

$$\Delta H^*(\sigma,T) = \Delta H_0^* - V^*\sigma$$

$V^*$ has the dimensions of volume and is usually called the *volume of activation*. The stress may be studied experimentally as a function of strain rate at constant temperature and also as a function of temperature at constant strain rate. Consequently the variables

$$[\sigma(\dot{\epsilon})]_{T=\text{const}} \ [\sigma(T)]_{\dot{\epsilon}=\text{const}}$$

can be measured. Show that

$$\Delta H^*(T,\sigma) = -RT^2 \frac{\left[\dfrac{\partial \sigma}{\partial T}\right]_{\dot{\epsilon}}}{\left[\dfrac{\partial \sigma}{\partial \ln \dot{\epsilon}}\right]_T}$$

(Use the chain rule for partial derivatives.) Show that

$$V^* = RT \left[\frac{\partial \ln \dot{\epsilon}}{\partial \sigma}\right]_T$$

if entropy does not depend on stress.

4.10    The paramagnetic ions in a maser (*Microwave Amplification by Stimulated Emission of Radiation*) may exist, when a magnetic field is applied, in three states. If these states are labeled 1, 2, and 3, and energy differences (per mole) between them are $\Delta E_{32}$ and $\Delta E_{21}$; $E_3 > E_2 > E_1$. Suppose that $\Delta E_{21} = 0.86$ calorie per mole, and $\Delta E_{32} = 0.8$ calorie per mole. For one mole of ions at $1.2°K$ under the above conditions, assuming that the ions only exist in states 1, 2, and 3, find the density of the ions in moles in each state (use equation 4.8).

4.11    (*a*) Compare the heat evolved in burning: (1) diamond, (2) graphite.

(*b*) Find $\Delta H$ for fabricating diamond from graphite; is it absorbed or evolved in the process?

(*c*) How is it possible to fabricate industrial diamonds?

4.12    (*a*) Draw a potential energy diagram for an uncatalyzed reaction.

(*b*) Compare (*a*) with a potential energy diagram for a catalyzed reaction. In each case indicate $\Delta H$ for the activated complex.

4.13    With the aid of a potential energy diagram (*PE* versus reaction coordinates) indicate the effect of a catalyst on a chemical reaction and its reverse.

4.14    (*a*) Draw an energy diagram of a spontaneous exothermic reaction indicating:

(1) enthalpy of reactants
(2) activation energy
(3) the activated state
(4) $\Delta H$ of the reaction
(5) the enthalpy of the products

(*b*) Do the same for an endothermic reaction.

4.15    (*a*) Draw an enthalpy diagram for the reaction of magnesium and oxygen. Label the kilocalories for each of the component and total reactions.

(*b*) Write the overall equation for each reaction.

(*c*) What reaction occurs at room temperature?

(*d*) In which reaction do we need to add heat initially?

CHAPTER FIVE

# Diffusion

SUMMARY

Matter may be transported through solids by diffusion. Diffusion, in thermodynamic terms, erases free energy gradients; on the atomic scale, it is the net effect of random atomic motions. In all cases it is activated by thermal energy. The mathematical formulation of the manner in which concentration differences disappear is expressed in *Fick's First* and *Second Laws*. Material may be transported by diffusive motion along surfaces, grain boundaries, and through the volume of solids. The fundamental mechanism by which atoms move through crystals depends on crystal structure, atomic sizes, and the extent of the defects in the crystals. Direct analysis of diffusivity and of phase diagrams is possible by the use of diffusion couples.

## 5.1 INTRODUCTION

Diffusion is the mechanism by which matter is transported through matter. Because the movement of each individual atom or particle is always obstructed by neighboring atoms or particles, its motion is an apparently aimless series of flights and collisions; however, the net result of a large number of these events can be an overall specific displacement of matter. The forces responsible for this motion may always be analyzed thermodynamically. An important feature of diffusive processes is that they are irreversible and therefore increase entropy.

The random molecular motion in fluids—liquids and gases—causes a relatively rapid disappearance of differences in concentration. In most solids, and particularly in crystalline ones, the atoms are more tightly bound to their equilibrium positions. However, there still remains an element of uncertainty caused by the thermal vibrations occurring in a solid which permits some atoms to move through the lattice at random. A large number of such movements results in a significant transport of material. This phenomenon is called *solid-state diffusion*. Even in a pure substance a particular atom does not remain at one equilibrium site indefinitely; rather, it moves from place to place in the material. In a pure material such movement is known as *self-diffusion*, and may be detected experimentally by radioactive tracers. In a mixture of more than one component, such as a binary metallic alloy, the process of *interdiffusion*, that is diffusion of one component through the lattice of the other, occurs. The various types of diffusion are important in all solid-state reactions and transformations to be considered in Chapters 6, 7, and 8.

## 5.2   THERMODYNAMICS OF DIFFUSION

Because diffusion occurs spontaneously, it should be viewed as a process which decreases free energy or alternatively, increases entropy. Increase of entropy is usually more apparent. Consider for instance, interdiffusion of components A and B, where complete solid solubility occurs in the A-B system. Blocks of A and B are placed in contact and heated to a temperature where diffusion readily occurs. The equilibrium state is a single homogeneous solid solution; presumably B will diffuse into the A, and vice versa, until equilibrium is reached. The irreversibility of the process is clear. Suppose a block of A were used in a similar experiment, with a block of A\*, which is an isotope of A. Interdiffusion again occurs and *chemical* effects are now absent; the most obvious change is the increase in entropy due to the mixing of the isotopes.

It is tempting to think of diffusion as "the great randomizer," as the process which wipes out all differences in concentration by random atomic motion. Such a description is incomplete and sometimes plainly incorrect. The eutectoid reaction, for instance, involves the breaking-up of a single, homogeneous, solid solution into two separate phases, sometimes of widely differing compositions. The diffusion which occurs during the above transformation creates, rather than removes, differences in concentration. In such a case, it is easier to think of diffusion as a reducer of free energy. This point may be proved by causing free energy gradients in a system through the application of electrical or magnetic fields or mechanical stresses.

## 5.3   MATHEMATICAL ANALYSIS OF DIFFUSION

The simplest possible diffusion system is illustrated in Figure 5.1. In this picture the flux of diffusing species $J_x$ is positive from left to right as the diffusing species moves from an initial high concentration $C_s$ to a lower concentration $C_x$, over a distance $\Delta x$ under *steady state* conditions. Flux is defined as the amount of material passing through a unit area perpendicular to the flux direction (flux is a vector) per unit time. In this case the concentrations $C_s$ and $C_x$ are constant, the concentration gradient $dc/dx$ is constant, and since $C_s > C_x$, the concentration gradient is negative from left to right. The amount of material passing through the slab in Figure 5.1 increases with increasing area $A$ and with increasingly negative gradient $dC/dx$. The coefficient of

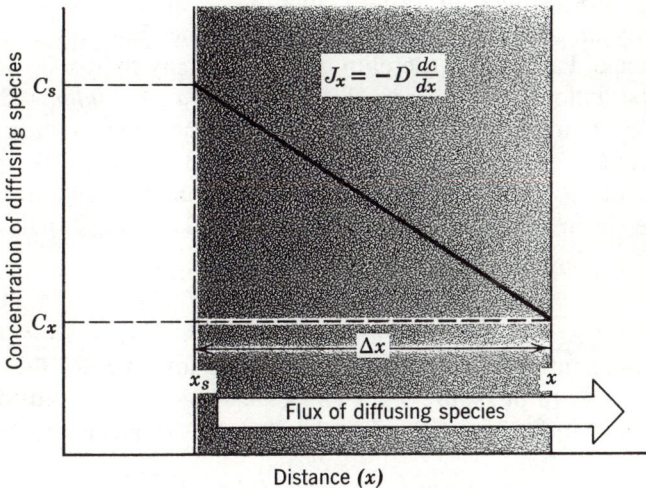

$$J_x = -D\frac{dc}{dx}$$

Figure 5.1    Steady-state diffusion.

proportionality for this system is analogous to that of electrical conductivity in the flow of electric current and is known as the *diffusivity* or *diffusion coefficient D*.   The behavior of the flux just described is given by the equation

$$J_x = -D\frac{dC}{dx} \tag{5.1}$$

This expression has come to be known as *Fick's First Law.*

A more common instance of diffusion than equation 5.1 arises when the concentration of the diffusing species (e.g., $C_x$ in Figure 5.1) changes with time.   This is true in the diffusion couple example of Section 5.2 and in many other situations when the diffusing species accumulates within the volume.   Under these "transient" or "unsteady-state" conditions, the gradient $dC/dx$ and, therefore, the flux $J_x$ in equation 5.1 change as time passes.   This can be represented by

$$\frac{dC_x}{dt} = \frac{d}{dx}\left[D\frac{dC_x}{dx}\right] \tag{5.2}$$

This equation has become known as *Fick's Second Law.*   Its derivation is illustrated in Problem 5.1.

Equations similar in form to equation 5.2 are encountered frequently in practical problems concerning the transport of heat and mass. Particularly useful in describing many solid-state diffusion situations, is the case of *diffusion in a semi-infinite solid,* illustrated in Figure 5.2.   Here concentration of diffusing species $C_x$ varies with distance $x$, time $t$, and diffusivity $D$.   If $D$ does not depend upon concentration (an assumption that must be proved in any practical situation), equation 5.2 may be simplified to

$$\frac{dC_x}{dt} = D\frac{d^2C_x}{dx^2} \tag{5.3}$$

This differential equation has been solved many times for the conditions shown in Figure 5.2.   The general method of solution is illustrated in Problem 5.2.   The equation that results is

$$\frac{C_x - C_0}{C_s - C_0} = 1 - \varphi\left(\frac{x}{2\sqrt{Dt}}\right) \tag{5.4}$$

Solid occupies   $0 < x < \infty$
$-\infty < y < \infty$
$-\infty < z < \infty$

Concentration

$C_s$

$C_0$

Initially

Later

Much later

$t = 0$   Out to $\infty$

$t = \dfrac{1}{36D}$   To $\infty$

$t = \dfrac{1}{D}$   To $\infty$

$t = 0$
$x = 0$

$x \longrightarrow$

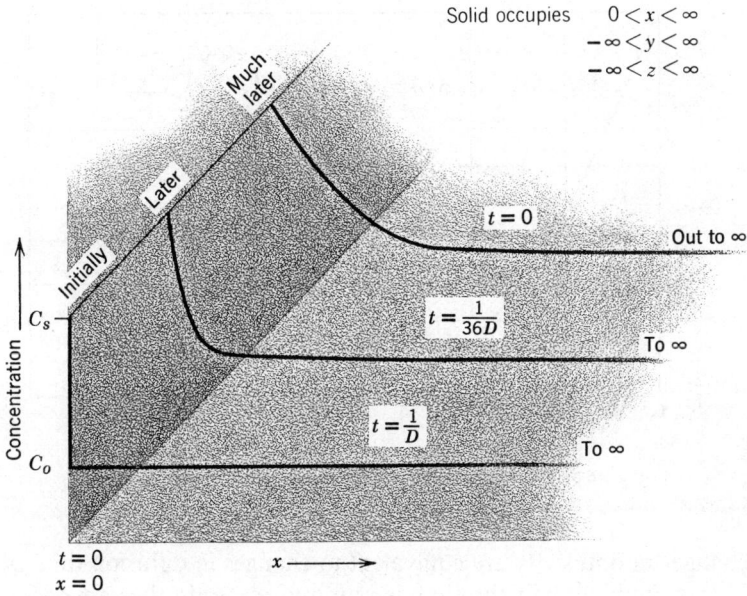

Boundary conditions:
$\left\{ \begin{array}{l} C = C_0 \text{ at } t = 0,\ 0 < x < \infty \\ C = C_s \text{ at } x = 0,\ 0 < t < \infty \end{array} \right\}$
i.e., the surface is maintained at concentration $C_s$.

Figure 5.2   Diffusion in a semi-infinite solid.

in which the function $\varphi(y)$ is the *normalized probability integral* or *Gaussian Error Function*. The values of $\varphi(y)$, as well as those of other common mathematical functions, are available in mathematical tables. By using equation 5.4 it is possible to evaluate quantitatively the information in Figure 5.2, on a single master plot. This is done in Figure 5.3, which is a graph of $1 - \varphi(y)$ versus $y$ converted to the terms in equation 5.4. The importance of the curve shown in the figure lies in the interrelation it demonstrated between distance, time, diffusivity, and concentration during diffusion. When $C_0$ and $C_s$ are known in a material of known $D$, $C_x$ must be a function of $x/\sqrt{Dt}$. For example, if it is desirable for the penetration depth to be doubled, the diffusion time must be four times as long; for $n$ times the original depth, the diffusion time must be multiplied by $n^2$. It should also be noted that

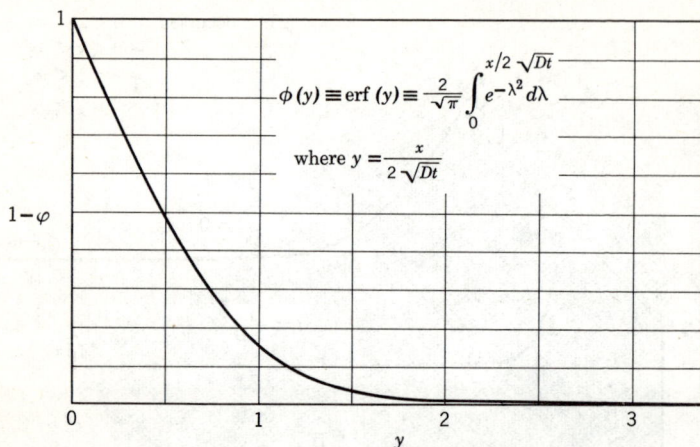

Figure 5.3    Use of the Gaussian error function for diffusion.

changes in diffusivity are equivalent to changes in diffusion time—
if $D$ is doubled, half the time is required to attain the same case
depth.  Equivalent degrees of saturation of objects of similar
shape can be calculated in the same way.  To bring some point
(e.g., the center) to a given concentration, or to diffuse in a given
fraction of the amount needed for total saturation, it is necessary
only to keep the same value of $L/\sqrt{Dt}$, where $L$ is a dimension
which characterizes the size of the object.  The application of the
principles of geometric similarity permits considerable liberty in
choosing $L$.  For example, if we wish to obtain the same average
concentration in a series of cubes of various sizes, we need only
maintain $L/\sqrt{Dt}$ constant for all the cubes, where $L$ is the length
of the cube edge and $t$ the diffusion time for each cube.  In the
same way, $L$ could be the thickness of very long wide sheets or
one side of a series of geometrically similar parallelepipeds.  It is
clear, then, that $L/\sqrt{Dt}$ is a powerful tool in generalizing the
principles by which diffusion operates to specimens of any size
and also in allowing for changes in the diffusivity of materials.

Although these solutions involve the assumption of constant
diffusivity, exact solutions for equation 5.3, in which $D$ varies with
concentration, are also available.  Complex diffusion problems
have also been solved by the use of computers.

## 5.4    EXPERIMENTAL MEASUREMENT OF DIFFUSION

The concepts developed in Section 5.3 serve as an outline for handling experimental diffusion data.  Most measurements have been made by placing two blocks of material in intimate contact —making them a *diffusion couple*—and measuring composition, distance, time, and temperature.  The proper preparation of the specimen and the accurate measurement of composition on a sufficiently fine scale present the major problems in utilizing the diffusion couple technique.

A diffusion couple can be made by actually placing two dissimilar blocks in contact or, in the case of metals, by electroplating one on the other.  The specimen can be cut on a plane like that shown in Figure 5.2.  Composition can then be measured by fine-scale x-ray analysis, hardness tests, or, if the diffusing species is radioactive, by one of several radiation counting techniques.  In the case of carbon diffusion in iron, for instance, the microstructure can sometimes be used as an indication of carbon content. An alternative procedure is to section the couple parallel to the surface shown in Figure 5.2 and to analyze it by using x-rays, mass spectrometry or by "wet" chemical analysis of chips that have been removed.

The result of such a composition measurement is the curve shown in Figure 5.2.  When this type of measurement is repeated at various temperatures, it is possible to find the diffusivity $D$ at these temperatures from the relationship shown in Figure 5.3. The diffusivity of many materials is found to obey the Arrhenius Equation (equation 4.1):

$$D = D_0 e^{-Q/RT} \tag{5.5}$$

The $D_0$ and $Q$ terms may be evaluated by the method described in Section 4.2.  Equations 5.4 and 5.5, together, specify the diffusion process completely.  The effects of time, distance, and temperature may be evaluated in practice by the methods shown in Section 5.3.

Until now, only the macroscopic aspects of diffusion have been considered.  A knowledge of the number of types of diffusion, and the possible atomic mechanisms for them, assist in understanding the phenomenon as a whole.

## 5.5    TYPES OF DIFFUSION

In general, the kinetic barrier to the movement of an atom through a solid lattice is greater than that to movement through a liquid or a gas.   This is reflected in the higher activation enthalpy or energy $Q$ that is necessary for *volume diffusion* through a solid than through a liquid or gas.   Volume diffusion is clearly important for a case like the one demonstrated in Figure 5.2.   If a grain boundary were present in the $x$ direction in Figure 5.2, a diffusing atom might be expected to move more easily down it than through the volume of the lattice, because the grain boundary is a higher energy region than the lattice.   Its presence therefore reduces the amount of activation energy which must come from other sources, before the atom will diffuse.   Thus the activation energy for *grain boundary diffusion* is lower than for volume diffusion.   If a crack existed in the $x$ direction in Figure 5.2, atoms would be transported inward by *surface diffusion*.   These three types of diffusion are illustrated schematically in Figure 5.4.   Here an equal-concentration profile is drawn for material diffusing to the right by volume, grain boundary, and surface diffusion.   The amount of energy necessary for each of these types relative to one another can be summarized by

$$Q_{vol.} > Q_{gb} > Q_{surf.} \qquad (5.6)$$

In the very few systems in which all three activation enthalpies have been determined, their relative values are

$$Q_{vol.} : Q_{gb} : Q_{surf.} \cong 4 : 3 : 2 \quad \text{or} \quad 4 : 2 : 1 \qquad (5.7)$$

Simultaneously typical values of $D_0$ are

$$D_{0\ vol.} > D_{0\ gb} > D_{0\ surf.} \qquad (5.8)$$

and range between 0.1 and 1.0 cm$^2$/sec.   The three coefficients of diffusion may be arranged:

$$D_{surf.} > D_{gb} > D_{vol.} \qquad (5.9)$$

for temperature of interest in solid-state reactions.   Figure 5.5 summarizes typical data for silver diffusion as a function of temperature.   At exceedingly high temperatures the inequality in equation 5.9 may be reversed; but this is of no practical signifi-

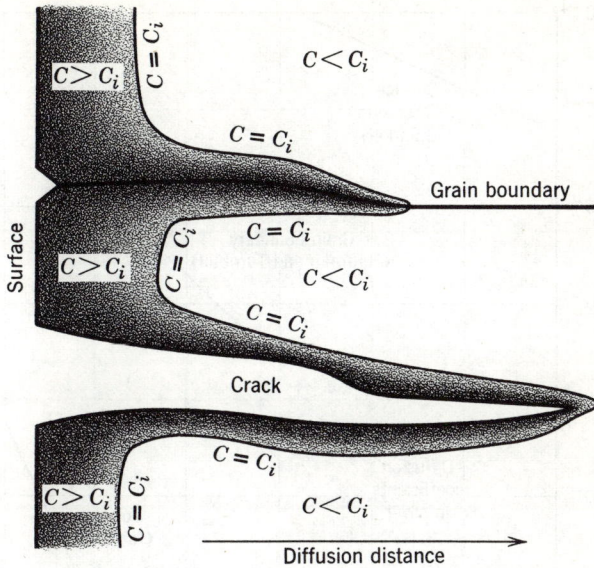

Figure 5.4    Equal-concentration profile of surface, grain boundary, and bulk diffusion in the same solid.

cance as in silver, for example, the temperature lies well above the melting point.

The relative importance of the three types of diffusion in actual processes does not depend on the diffusion coefficient alone. The amount of material transported by any of the three types of diffusion is given by equation 5.1 (Fick's First Law). For the same composition gradient, the amount of material transported depends also on the effective area through which atoms diffuse. If the effective "thickness" of a grain boundary or of a surface diffusion path is assumed to be several atom spacings ($\sim 10^{-7}$ cm), the areas of these paths are very small compared to those of the volume diffusion path. The data shown in Figure 5.5 suggest that only at low temperatures, when the value of $D_{\text{vol.}}$ has dropped far below that of the others, do grain boundaries and surfaces become really important as mass transport paths. It is also possible to show that grain boundary diffusion competes with volume diffusion only in very fine-grained material, depending on the ratio of

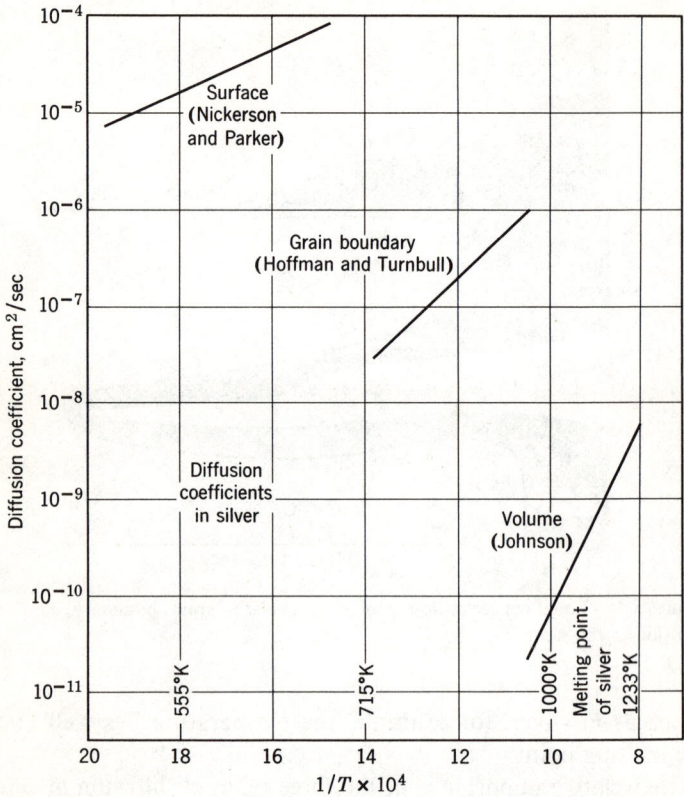

Figure 5.5    Diffusion coefficients in silver.

$D_{gb}$ to $D_{vol.}$. In processes such as sintering and oxidation, grain boundary diffusion and surface diffusion are very important. The remainder of this chapter, however, will be devoted to volume diffusion, which is the type most frequently encountered.

## 5.6    ATOMIC MECHANISMS OF VOLUME DIFFUSION

The mechanics of the motion of atoms or molecules in fluids are relatively well known. In most gases the molecules move in straight paths until they collide. The change in path resulting

from these collisions may be estimated.  The cumulative motion
of these molecules may then be predicted from the collision
parameters, and this procedure yields the diffusive properties of
the gas in question.  The mechanism of diffusion in solids is less
clear.  Somehow an atom in a certain lattice site is transferred to
an adjacent site.  This is the basic step in the diffusion process.
The actual method by which this transfer occurs, however, has
not been unambiguously demonstrated.  Figure 5.6 illustrates
some of the possibilities.  Figure 5.6*a* shows an atom moving to
the next lattice site and occupying a *vacancy* there.  In Figure
5.6*b* an atom moves out of its lattice and becomes an *interstitial
atom* which is free to move.  In Figure 5.6*c* atoms in a *ring* simultaneously move to adjacent lattice sites.  In Figure 5.6*d* two atoms
change places directly.

   The vacancy mechanism currently appears to be the most

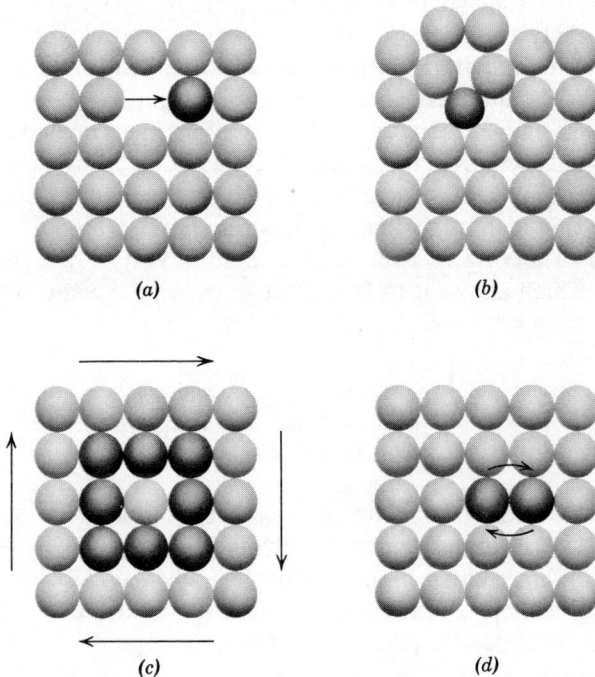

Figure 5.6   Atomic mechanisms for diffusion in solids.

probable one in self-diffusion and in diffusion of substitutional solid solution elements and ions in metals and ceramics. If vacancies are already present, the activation energy for diffusion in this kind of situation is only that required for an atom to part with one set of near neighbors and move into a vacant site among another set. Reasonable agreement has been found between observed diffusivities and calculations based on this model.

The interstitial mechanism is important in two cases. A solute atom, when small enough to dissolve interstitially, moves most readily by this mechanism. This occurs particularly when carbon, nitrogen, oxygen, and hydrogen dissolve and diffuse in metals, and ions of alkali metals and various gases in silicate glasses and vitreous materials. In these cases, diffusivities are quite high at surprisingly low temperatures. The diffusion of atoms sufficiently small to be present interstitially in the lattice greatly affects the mechanical behavior of metals. Interstitial diffusion also occurs in materials subjected to neutron irradiation. High-energy neutrons supply the energy necessary to "knock out" an atom from a lattice site into an interstitial position from which it is then free to move. In all other substitutional solid solutions, however, the amount of energy required to place an atom in an interstitial site, by thermal excitation, is too great for interstitial diffusion to occur at a significant rate. The direct interchange mechanism shown in Figure 5.6d is unlikely because it too requires a high activation energy. The ring mechanism has a sufficiently low activation energy which allows it to be operative in some situations and to explain some experimental observations.

## 5.7    DIFFUSION AND DEFECTS:
### THE HARTLEY-KIRKENDALL EFFECT

If a binary diffusion couple is assembled of materials A and B, and B diffuses much more rapidly through A than vice versa, complications result. Figure 5.7 depicts schematically what happens. The great difference in diffusion rate results in actual mass transport across the original A-B interface, and the interface markers appear to "migrate." The mass transport may proceed with such rapidity that voids are left on the B-side, as in Figure

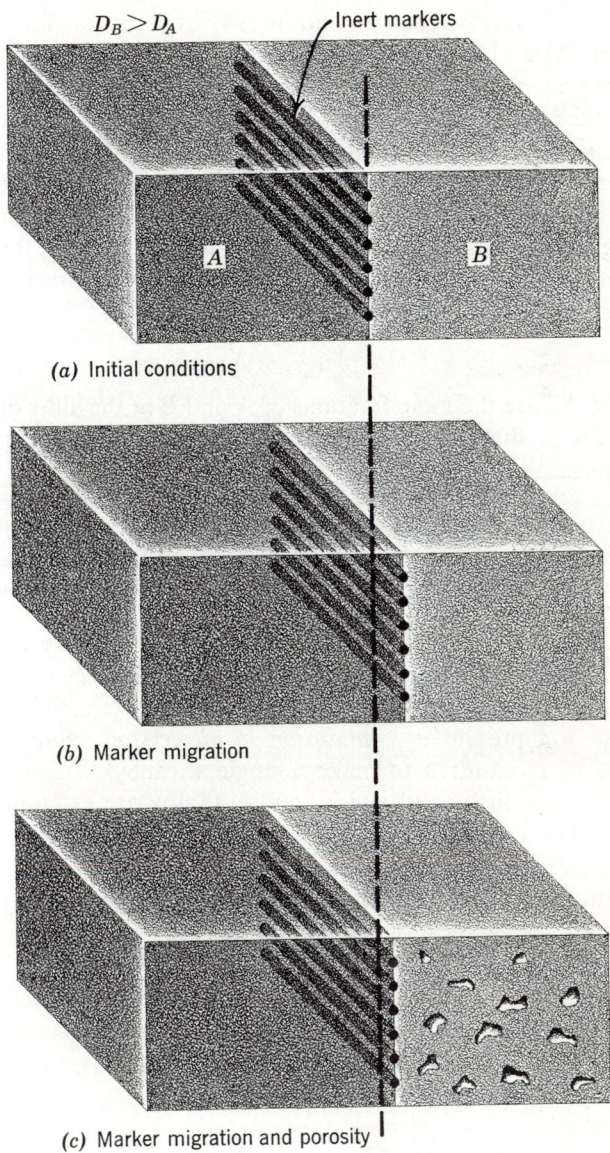

$D_B > D_A$

Inert markers

$A$

$B$

(a) Initial conditions

(b) Marker migration

(c) Marker migration and porosity

Figure 5.7    The Hartley-Kirkendall Effect.

5.7c.  This effect, observed in 1946 by Hartley, who used cellulose acetate (A) and acetone (B), and in 1947 by Kirkendall, who used copper (A) and brass (B), is generally viewed as evidence for the vacancy mechanism.  Both the mass transport and the voids of Figure 5.7 may be accounted for easily by a high flux of vacancies to the right, that is, the rapidity of motion of B to the left.

In addition, the *Hartley-Kirkendall Effect* shows that interdiffusion in binary alloys, for which we have previously accounted by a single coefficient $D$, consists of two classes of motion, that of A atoms and that of B atoms.  Analysis in detail by Darken in 1948 shows that

$$D = X_B D_A + X_A D_B \qquad (5.10)$$

$X_B$ and $X_A$ are the mole fractions of A and B in the alloy of interest.  $D_A$ is the diffusion coefficient of B in pure A, and $D_B$, vice versa.

## 5.8    VACANCY EQUILIBRIUM IN CRYSTALS

Although mass transport through crystals may result from mechanisms other than vacancy motion, one fact makes the vacancy mechanism particularly attractive: vacancies, unlike other defects, are present *at equilibrium in all crystals*.  Suppose the energy $E_v$ is required to make a single vacancy.  In solids, $PV$ changes very little, and energy and enthalpy are essentially the same.  The entropy of the "mixture" of full and vacant lattice sites in a crystal is calculated in exactly the same manner as the configurational entropy of a binary alloy.  (See Problems, Chapter 2.)

For $N_v$ vacant sites, $N_f$ filled ones, totaling $N_0$,

$$S_v = k \ln \left\{ \frac{[N_0 !]}{[N_v !][N_f !]} \right\} \qquad (5.11)$$

Using the Stirling approximation, and substituting $N_f = N_0 - N_v$,

$$S_v = k\{N_0 \ln N_0 - (N_0 - N_v) \ln (N_0 - N_v) - N_v \ln N_v\} \qquad (5.12)$$

Now we write the free energy:

$$F_v = N_v E_v - T S_v \qquad (5.13a)$$

and

$$\frac{\partial F_v}{\partial N_v} = 0 \qquad (5.13b)$$

for the equilibrium criterion. The result is

$$\frac{N_v}{N_0 - N_v} = e^{-E_v/kT} \approx \frac{N_v}{N_0} \qquad (5.14)$$

If, for instance, $E_v$ is 20,000 cal/mole, at $1000°K$, one site in $10^5$ will be vacant.

## 5.9  DIFFUSION AND DEFECTS: OXIDES AND IONIC CRYSTALS

Oxygen diffuses through many oxides by vacancy migration. However, the vacancies involved are usually far more plentiful than the equilibrium number discussed in Section 5.8. The best source of oxygen vacancies in oxides are the dissolved impurities. For instance, if a divalent or trivalent oxide is dissolved as an impurity in a tetravalent oxide lattice, the cation sites will all be occupied, and some anion sites left vacant. Such a vacancy concentration is independent of temperature, in contrast to the discussion of Section 5.8. Very pure oxides, therefore, may have very low diffusivities for oxygen. Cations may diffuse through oxides by vacancy migration or, because of their relatively small size, by interstitial motion. Once again: the majority of vacancies and interstitials (this time in the cation lattice), are chemically induced. Usually an excess or depletion of cations exists because the atmosphere under which the oxide was made was either too strongly oxidizing or reducing. Figure 5.8 plots the variation (with temperature) of various oxide diffusivities.

In pure ionic crystals, defects of the *Frenkel* type (cation vacancies and interstitial cations) and of the *Schottky* type (cation and anion vacancies) are both observed. Such defects preserve local electrical neutrality and exist at equilibrium in the same manner as the simple vacancies of Section 5.8. Which type of defect prevails depends on whether, for the particular material, interstitial cations or anion vacancies take more energy to produce. In addi-

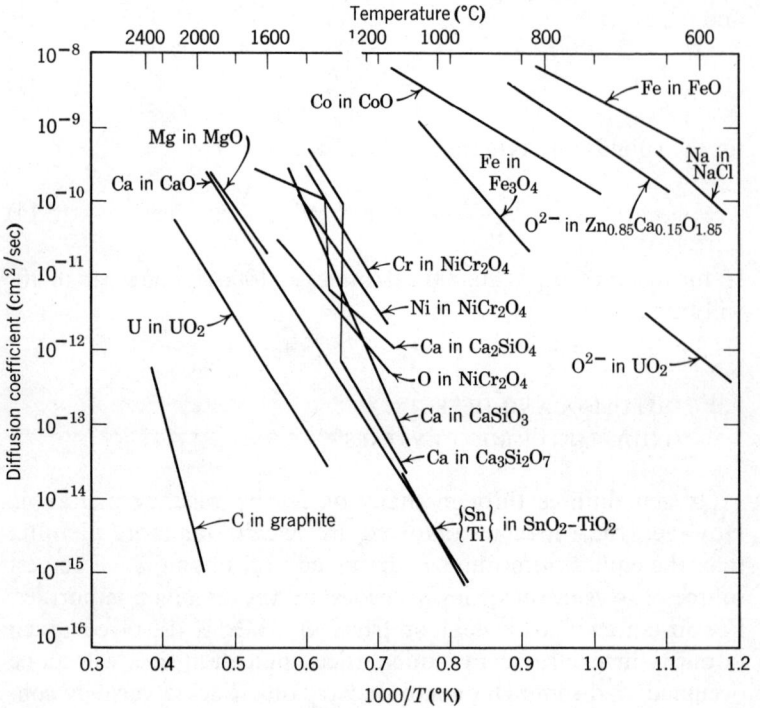

Figure 5.8  Diffusion coefficients in some crystalline oxides (from several sources).
Figure 8.9 from *Introduction to Ceramics* by W. D. Kingery (Wiley, 1960), p. 232.

tion there is the usual impurity effect: a divalent cation, dissolved in a lattice of monovalent cations, requires, for electrical neutrality, that a defect of unit negative charge exist somewhere nearby. The electrical conductivity at high temperature for ionic crystals is due almost wholly to diffusion of ions rather than electrons. Consequently, the electrical conductivity in such cases is related to the diffusion coefficient:

$$\sigma = (\text{const.}) \left[ \frac{c_d q_d^2}{kT} \right] D_d \qquad (5.15)$$

$c_d$ and $q_d$ are the concentration and charge, respectively, of the defect involved. The constant depends on the defect; it is unity for interstitial ions and somewhat more than unity for vacancies.

## 5.10    DIFFUSION IN AMORPHOUS AND VITREOUS SOLIDS

In amorphous and vitreous solids the mechanism of diffusion is likely to depend on the particular details of the bonding and structure. For instance, polymeric materials are characterized by very strong (covalent) primary bonds which hold the molecules together and by much weaker secondary bonds between the molecules. Diffusion in polymers occurs by motion of entire molecules, which is analogous to the interstitial mechanism in crystalline materials. If the polymer chain is long and bulky, motion becomes more difficult; polymer diffusion rate measurements are often used in this connection to determine the molecular weight of the polymer. In vitreous or glassy solids, many species take advantage of the irregularities of the structure. In silicate glasses, silicon ions are very strongly bonded to oxygen and diffusivity of silicon in such materials is extremely low. However, the irregular silicate network contains a large number of sites for positive ions, which usually fill only a small fraction of the sites. Whereas the silicon ions are usually at the center of the silicate tetrahedra, surrounded by oxygen ions, positive ions of the alkali metals sit in peripheral positions held by weaker (coulomb) bonds. Consequently, sodium, potassium and similar elements diffuse through silicate glasses with ease. Also, the holes in the silicate network allow small atoms such as hydrogen or helium to diffuse with great rapidity. This transparency to hydrogen and helium may limit the use of glass for extremely high vacuum applications (less than $10^{-10}$ mm Hg).

## 5.11    DIFFUSION COUPLES AND THE PHASE DIAGRAM

Thus far, the discussion of diffusion has assumed that both components are completely soluble in each other. Obviously this is not always the case. Many diffusion couples have compositions which fall in two-phase regions. This is of particular interest, since we know that the compositions of the two phases will be fixed, if the couple is truly binary. If compositions are constant, no composition gradients exist, and diffusion cannot occur in two-phase regions. Consequently, two-phase regions will not grow by diffusion in a binary diffusion couple. Only one-phase regions will be observed; here composition gradients occur and composition may vary between the limits of solubility. Two typical cases are shown in Figures 5.9 and 5.10 for diffusion couples which consist, initially, of the pure components. Such behavior makes the diffusion couple a powerful tool for the investigation of phase diagrams. Figure 5.11 (see page 94) shows the microstructure and composition vs. distance plot for a tungsten-iridium diffusion

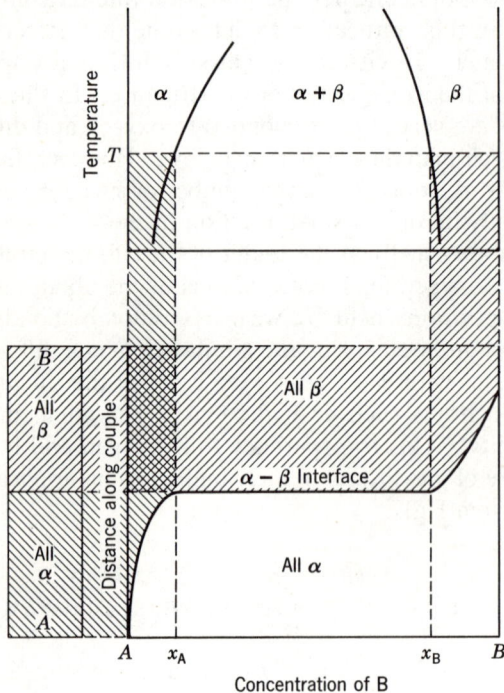

Figure 5.9   Diffusion couple with limited solubility.

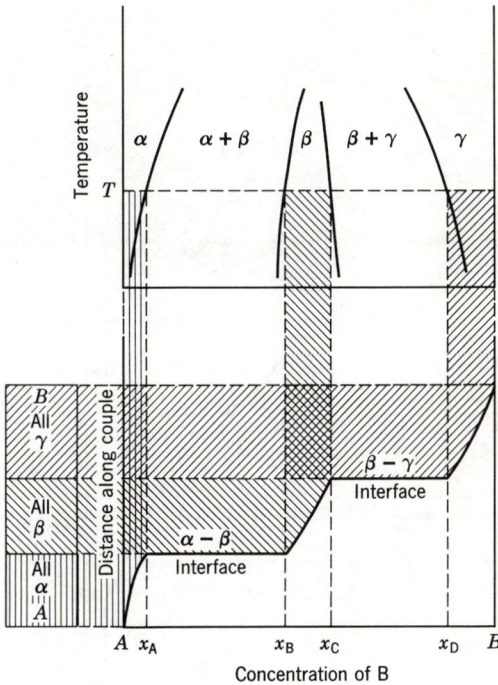

Figure 5.10   Diffusion couple with limited solubility and intermediate phases.

couple. In addition to the terminal solid solutions (W in Ir, Ir in W), there are two intermediate phases in the W-Ir system.

## DEFINITIONS

*Diffusion.* The motion of matter through matter.

*Diffusion Coefficient.* The constant of proportionality in both of Fick's Laws.

*Diffusion Couple.* An assembly of two materials in such intimate contact that each diffuses into the other.

*Error Function.* A mathematical function found in some solutions of Fick's Second Law, existing by agreed definition (much as do sines and cosines).

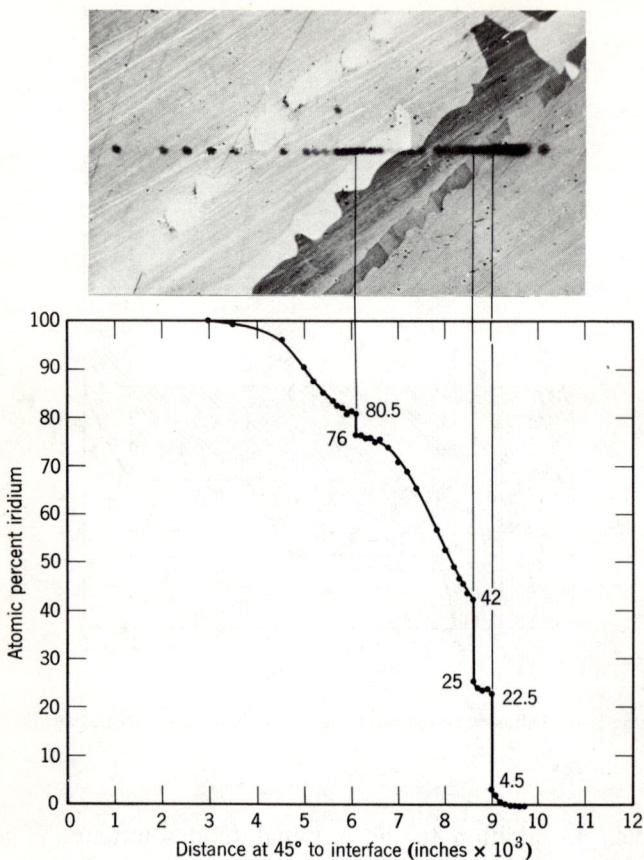

Figure 5.11    Tungsten/iridium diffusion couple 1955°C for 117 hours. (Courtesy of E. J. Rapperport.)

*Fick's Second Law.*    The rate of change of composition is proportional to the second derivative (*Laplacian*) of the concentration.

*Fick's First Law.*    The flux of a diffusing species is proportional to the concentration gradient.

*Grain Boundary Diffusion.*    Atomic migration along the grain boundaries.

*Interstitial.*    An atom located in the space between lattice sites.

*Self-diffusion.*    The migration of atoms in pure materials.

*Substitutional.*    An impurity atom which has replaced an atom on a lattice site.

*Surface Diffusion.*   Atomic migration along the surface of a phase: for instance, along a solid-vapor interface.

*Volume Diffusion.*   Atomic migration through the bulk of the material.

## BIBLIOGRAPHY

Cottrell, A. H., *Theoretical Structural Metallurgy,* St. Martin's, New York, 1957. Chapter 12 treats metallic diffusion in a clear, elementary manner.

Darken, L. S., and R. W. Gurry, *Physical Chemistry of Metals,* McGraw-Hill, 1953. Chapter 18 is a graduate-level account.

Kingery, W. D., *Introduction to Ceramics,* Wiley, New York, 1960.   Chapter 8 is an elementary treatment of diffusion in ceramics.

Le Claire, A. D., "Diffusion in Metals," p. 265 of *Progress in Metal Physics,* vol. IV, Pergamon Press, 1953.

Le Claire, A. D., "Diffusion of Metals in Metals," p. 306 of *Progress in Metal Physics,* vol. I, Butterworth, 1949.   Two comprehensive, advanced articles on diffusion in metals.

Shewmon, P. G., *Diffusion in Solids,* McGraw-Hill, 1963.   Extensive coverage of the entire subject on a senior and first-year graduate student level.

## Problems

5.1   Derive Fick's Second Law (equation 5.2) by recognizing that the time rate of composition change in the unit area slab of Figure 5.1 is equal to the difference between the flux into the slab, $J(x)$, and the flux out of the slab, $J(x + \Delta x)$; also, that $J(x + \Delta x)$ equals $J(x) + (\partial J/\partial x)\Delta x$.

5.2   Derive the solution to Fick's Second Law.   The solution will have the form of equation 5.4.   If the solid extends to infinity from $x = 0$, the appropriate boundary conditions will be $c = c_s$ at $x = 0$ and all values of time, $c = c_0$ at $t = 0$ for all values of $x$.   Also, at all values of time, $c = c_0$ at $x = \infty$.   The solution will be facilitated by changing the variables in equation 5.2 so that $\lambda = x/\sqrt{t}$ and then $y = x/2\sqrt{Dt}$.

5.3   The diffusion rate of carbon in $\alpha$ iron (ferrite) and $\gamma$ iron (austenite) are given by

$$D_\alpha = 0.0079 \text{ cm}^2/\text{sec } e^{(-18.100 \text{ cal/mole})/RT}$$
$$D_\gamma = 0.21 \text{ cm}^2/\text{sec } e^{(-33,800 \text{ cal/mole})/RT}$$

(*a*)   Calculate the two diffusion coefficients at 800°C and at 1000°C.

(*b*)   Explain the magnitude of $D_\alpha$ compared to $D_\gamma$ in terms of atomic structure.

(*c*)   Draw a composition-distance plot of a carbon-iron diffusion couple heat-treated at 800°C.   (See iron-carbon phase diagram, Figure 7.7.)

(d) Why are commercial carburizing treatments of steel carried out when the steel is austenitic?

5.4   Calculate the carburizing time necessary to achieve a carbon content of 0.8% at a depth of 0.2 mm in a sample of steel which originally contained 0.2% carbon.  Assume that the carbon content of the surface can be maintained at 1.3%, and that the carburizing treatment is made at 950°C.  Use the $D\gamma$ data from Problem 5.3 and the table of erf($y$) below.

| $y$ | ERF $y$ | $y$ | ERF $y$ | $y$ | ERF $y$ | $y$ | ERF $y$ |
|-----|---------|-----|---------|-----|---------|-----|---------|
| 0   | 0       | 0.5 | 0.521   | 1.0 | 0.843   | 2.0 | 0.995   |
| 0.1 | 0.112   | 0.6 | 0.604   | 1.2 | 0.910   | 2.4 | 0.999   |
| 0.2 | 0.223   | 0.7 | 0.678   | 1.4 | 0.952   |     |         |
| 0.3 | 0.329   | 0.8 | 0.742   | 1.6 | 0.976   |     |         |
| 0.4 | 0.428   | 0.9 | 0.797   |     |         |     |         |

5.5   The decarburization of steel may be expressed by another solution of Fick's Second Law:

$$(C - C_s) = (C_0 - C_s)\, \mathrm{erf}\left(\frac{x}{2\sqrt{Dt}}\right)$$

If a 0.9% carbon steel is held 10 hours at 950°C in a decarburization atmosphere which maintains surface concentration at 0.1%, at what depth below the surface will the carbon content be 0.8% carbon?  (Use $D_\gamma$ from Problem 5.3 and erf($y$) values from Problem 5.4.)

5.6   During the solidification of a metal (Chapter 7) a structure may be produced in which the composition varies from point to point in the casting.  Suppose the mean concentration is $C_0$, and that the concentration varies as

$$(C - C_0) = \Delta C_{\mathrm{max.}} \sin\left[\frac{\pi x}{l}\right]$$

The point concentration varies sinusoidally with the maxima and minima, $C = C_0 \pm \Delta C_{\mathrm{max.}}$ a distance $l$ from each other.  In this case the solution to Fick's Second Law is:

$$\Delta C = \Delta C_{\mathrm{max.}} \sin\left[\frac{\pi x}{l}\right] e^{-\pi^2 Dt/l^2}$$

In a particular case where refractory metal contains 75% solute, there are local variations in composition from 85% to 65% over a typical distance of 0.001 cm.  How much time at 1500°C would be required before the composition would vary from 77% to 73% solute over the same distance?   Use $D = 1.3 \times 10^{-12}$ cm$^2$/sec.

5.7    Discuss the conditions under which a metastable phase could be found in a binary diffusion couple.

5.8    At what relative temperatures do surface, grain boundary, and volume diffusion predominate?   How do these compare with the temperatures at which the respective diffusivities are equal?

5.9    What effect would grain size have on the relative predominance of grain boundary or volume diffusion?   At what grain size would an equal amount of material be transported in silver at 500°C (Figure 5.5)? Assume grain boundary thickness is $5 \times 10^{-8}$ cm, and that grains are hexagonal in shape.   Use the data:

$$D_{GB} = 0.025 \text{ cm}^2/\text{sec } e^{-20,000/RT}$$
$$D_{vol.} = 0.895 \text{ cm}^2/\text{sec } e^{-49,950/RT}$$

5.10    What is the driving force for selfdiffusion?

5.11    How would you expect inert gases to diffuse through metals?

5.12    What might be the reason for heating a nickel tube containing Ni-Cu-Fe-Co-Al-Ti Alnico alloy compacts in a hydrogen atmosphere at sintering temperatures while the interior of the tube is attached to a vacuum system?

5.13    Discuss the relative merits of vacuum versus hydrogen sintering of (1) stainless steel, (2) niobium, and (3) molybdenum.

5.14    Describe briefly an example of (1) a diffusionless transformation, (2) a transformation involving diffusion.

5.15    Describe in detail how you would process electrolytic iron powder embrittled by hydrogen in order to make it more pressible.

5.16    (a) Describe, with the aid of simple sketches, an internal friction apparatus.
    (b) Show how diffusion of interstitials in BCC metal can be shown by an internal friction peak.
    (c) Why do interstitial atoms not give rise to an internal friction peak in FCC metals?

5.17    In a short essay with figures describe the use of radioactive tracers in diffusion experiments.

CHAPTER SIX

# Phase Changes

SUMMARY

Phase transformations begin with the appearance of a number of very
small particles of the new phase which then grow until the change of
phase is complete. Although the replacement of the old phase by the
new is accompanied by a decrease in free energy, the existence of a surface between the two phases increases free energy. Atomic mismatchings
at the surface, "broken" bonds, elastic and plastic strains also contribute
to this increase. The growth of the new particle therefore depends on the
ratio of surface area to volume; small particles will tend to redissolve, and
large ones tend to grow. There is, in fact, a critical size separating those
that redissolve, which are called *embryos,* from those that grow, which are
called *nuclei.* The free energy of a particle is at a maximum at the critical radius. It is the achievement of this free energy which is the energy
barrier in the kinetics of nucleation. Nucleation usually occurs heterogeneously at dislocations, grain boundaries, foreign bodies, or at any
other place where the surface energy increase necessary for the nucleation
process may be reduced or the magnitude of the volume free energy
change increased. It is possible to predict, from kinetic principles, the
temperature dependencies of nucleation and growth and therefore the
overall transformation rate. Growth proceeds by material transfer into
the growing nucleus until the transformation is complete.

## 6.1 INTRODUCTION AND KINETICS

We are now in a position to consider the means by which phase
changes occur. Suppose the temperature of a piece of pure,
single-phase material is changed so that a new phase is now stable.
How does the new phase appear and the old disappear? It would
be unrealistic to suppose that such a change occurs all at once by
simultaneous, cooperative motion of every atom in the material.

98

In fact, the entropy of activation for such a process (equation 4.6) would be negative and astronomical in magnitude. The phase change must begin on a very small scale. First, *nuclei* of the new phase appear, perhaps only a few hundred atoms in size. Second, the nuclei then grow in size until all the material is transformed. The two principal steps, *nucleation* and *growth*, may be broken down in many cases to substeps. Nucleation may involve: (*a*) the assembly of the proper kinds of atoms by diffusive or other types of motion; (*b*) structural change into one or more unstable intermediate structures; (*c*) formation of nuclei of the new phase. Each step is expected to have an activation energy. One step will usually proceed more slowly than the others and limit the overall rate of the process. The observed activation energy will then be that of the rate-limiting step. The system will pass through unstable intermediate states only if the overall reaction rate is increased by doing so.

Growth may also be broken down into steps. Typical of the growth steps are: (*a*) transfer of material by diffusion through the old phase; (*b*) transfer across the phase boundary into the new phase; (*c*) transfer into the interior of the new phase by diffusion. In many cases part (*b*) may be further subdivided. All the above steps are thermally activated processes; that is, barriers are encountered which are surmounted by thermal energy. The nature of these barriers is our next concern.

## 6.2    HOMOGENEOUS NUCLEATION

Suppose a small region of a new, stable phase appears in the middle of the old, unstable material. From the thermodynamics a free energy decrease $\Delta F_v$ per unit volume is expected, contributing to the stability of the region. However, the region is bounded by a surface, and such a surface (see Chapter 3) has, at least, a positive free energy $\gamma$ per unit area associated with it. Since nuclei are small the ratio of surface to volume is high, and surface energy is therefore quite important. Other problems may also occur. Volume changes may accompany the transformation, and the nuclei may therefore have long or short-range stress fields, and consequently a further increase in free energy. It is therefore apparent that all regions of the new phase may not be stable, and some may disappear rather than grow.

For simplicity let us consider the solidification of a pure material, thus eliminating the stress problem. Furthermore, suppose the nuclei are spherical, although the method discussed below may be applied to any shape. If the temperature is suddenly dropped below the melting point, the free energy change $\Delta F_v$ per unit volume solidified will be negative, amounting to $\frac{4}{3}\pi r^3\ \Delta F_v$ for a nucleus of radius $r$. The total free energy change for a particle of radius $r$ is

$$\Delta f = 4\pi r^2\gamma + \tfrac{4}{3}\pi r^3\ \Delta F_v \tag{6.1}$$

Each term in this equation is plotted in Figure 6.1. As the particle first increases in size, its free energy also increases until the

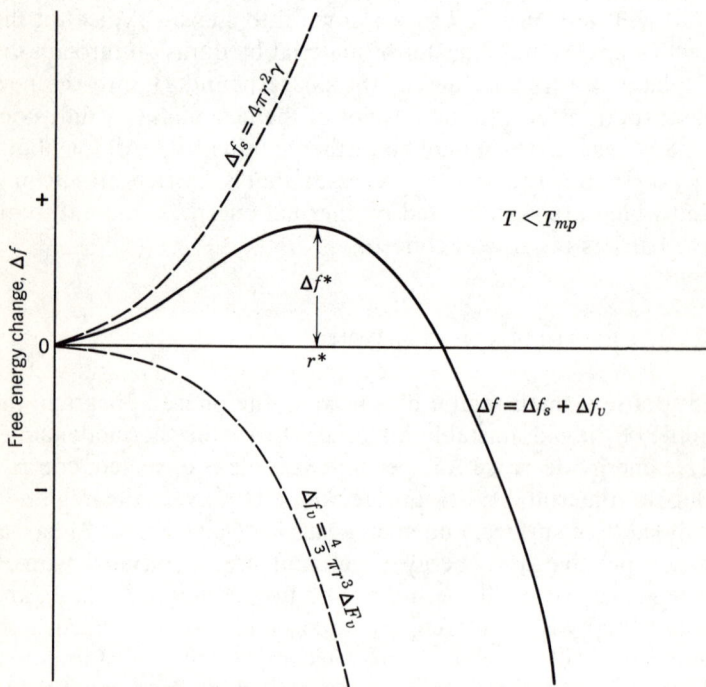

Figure 6.1  Nucleation of spherical solid nuclei from a pure liquid.

radius $r^*$ is reached.   Particles of radius less than $r^*$ will tend to redissolve, thus lowering the free energy.   However, particles of radius larger than $r^*$ will lower free energy by growing.   Subcritical particles ($r < r^*$) are called *embryos;* the opposite case ($r > r^*$), *nuclei.*   To form a nucleus, the energy $\Delta f^*$ must be added, and here is where thermal activation enters.   The radius of the nucleus is the reaction coordinate (Chapter 4), and the activation energy is $\Delta f^*$.   From Figure 6.1 it is clear that $\Delta f^*$ and $r^*$ may be calculated by maximizing equation 6.1.   By setting the first derivative equal to zero, we obtain

$$r^* = \frac{-2\gamma}{\Delta F_v} \tag{6.2a}$$

$$\Delta f^* = \frac{16\pi\gamma^3}{3(\Delta F_v)^2} \tag{6.2b}$$

The rate of nucleation will be determined by $\Delta f^*$, as described in Chapter 4.   A good approximation for the temperature variation of $\Delta f^*$ may be obtained if we recognize that $\Delta H_v$ and $\Delta S_v$ vary slowly with temperature.   At the melting point ($T_0$), $\Delta F_v = 0$, and

$$\Delta S_v(T_0) = \frac{\Delta H_v(T_0)}{T_0} \tag{6.3}$$

Then we write

$$\Delta F_v(T) \cong \Delta H_v(T_0) - T\,\Delta S_v(T_0)$$
$$\cong \Delta H_v(T_0)\left[1 - \frac{T}{T_0}\right] \tag{6.4}$$

which is quite good for temperatures close to $T_0$.   This is exactly what we are interested in.   The surface energy also is relatively insensitive to temperature, and therefore the approximate temperature variation of $\Delta f^*$ and $r^*$ is obtained by substituting equation 6.4 for $\Delta F_v$ in equations 6.2a and 6.2b.   The result is plotted in Figure 6.2.

If the total number of particles of the new solid phase is $n_0$, the number of nuclei $n^*$ may be calculated by equation 4.5:

$$n^* = n_0 e^{-\Delta f^*/kT} \tag{6.5}$$

The nuclei grow by addition of atoms, the final step in nucleation. The rate of addition is proportional to the frequency of motion ($\nu$)

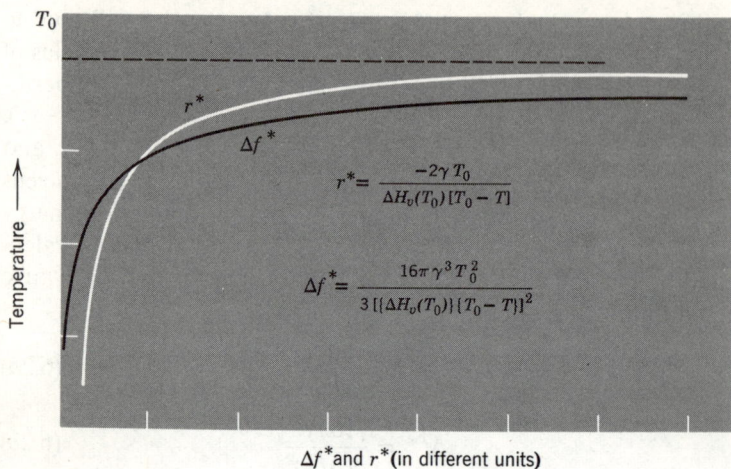

Figure 6.2  Temperature variation of $\Delta f^*$ and $r^*$ on the basis of equation 6.4. $T_0$ is the equilibrium transformation temperature.

into the nucleus for a single atom, multiplied by the number of atoms ($m^*$) which actually do so, that is, the immediate neighbors to the nucleus. Consequently,

$$\dot{N} = n^*vm^* \qquad (6.6)$$

where $\dot{N}$ is the rate of nucleation measured in units of nuclei per second. The motion of atoms into the nucleus is diffusive and therefore (Chapter 5)

$$v \propto e^{-\Delta f_D/kT} \qquad (6.7)$$

where $\Delta f_D$ is the activation energy for diffusion. $m^*$ is geometrical and depends on the shape and size of the nucleus. The previous three equations may be combined to give

$$N \propto m^*e^{-(\Delta f^* + \Delta f_D)/kT} \qquad (6.8)$$

for the temperature variation. Using equations 6.2b and 6.4 (or Figure 6.2) for $\Delta f^*$, the approximate temperature variation of the nucleation rate may be calculated. Figure 6.3 is the result. As temperature drops, $\Delta f^*$ falls violently and the nucleation rate increases. Soon, however, $\Delta f^*$ becomes negligible compared to

$\Delta f_D$ and $\Delta f_D$ thus dominates equation 6.8, and $\dot{N}$ then decreases with temperature. Consequently, there is a maximum in the homogeneous nucleation rate which may be considerably below the melting point.

So far, this section has dealt with nucleation of a solid phase from a liquid. For nucleation of a solid from vapor the same considerations apply, but the magnitudes of $\Delta F_v$ and $\gamma$ will generally be much larger. As Chapter 3 pointed out, the surface energy between the solid and vapor is roughly one order of magnitude larger than the solid-liquid surface energy. Also, addition of atoms to the nucleus may occur directly from the vapor; diffusion is not needed. If a solid phase is nucleated from a solid, the situation is more complicated. Diffusion in solids is relatively slow, and the nuclei grow slowly. In addition the change in volume due to the phase change leads to strains whose energy must be accounted for. Some of the strain may be localized near the boundary of the nucleus. The extra energy due to this strain may be added to the surface energy as it is also proportional to the

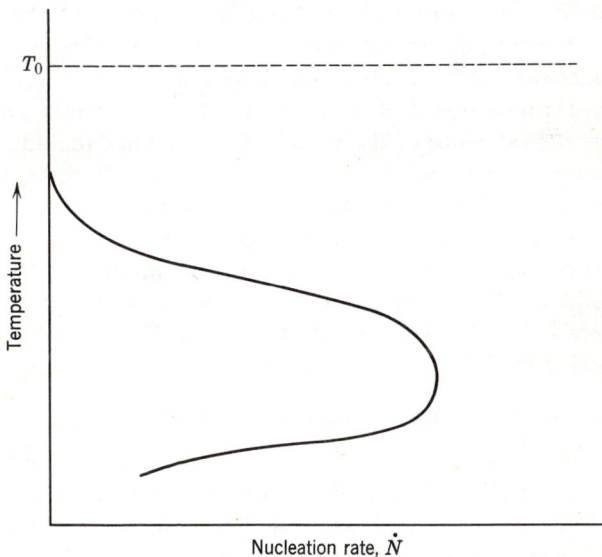

Figure 6.3    Homogeneous nucleation rate.

surface area of the nucleus. Some of the strain may be long range by nature and proportional to the volume of the nucleus. We may add this to $\Delta F_v$. Since $\Delta F_v$ is negative, its magnitude is thereby decreased. Generally solid-solid nucleation increases the magnitude of $\gamma$ and decreases the magnitude of $\Delta F_v$ because of strain energy. This means, according to equations 6.2a and 6.2b, larger $r^*$ and $\Delta f^*$ and consequently slower nucleation. For such reasons, and also slower diffusion rates (larger $\Delta f_D$), nucleation in solids is a very sluggish process, and supercooling and nonequilibrium structures are the rule rather than the exception in solid materials. In all solid-solid cases we may expect transformations to proceed at small or negligible rates near the equilibrium transformation temperatures, with some supercooling necessary to increase the rate.

### 6.3   HETEROGENEOUS NUCLEATION

We may now use the equations of Section 6.2 to predict supercooling by use of available free energy and surface energy data. The predictions will exceed any common observation. Equation 6.2b has given us an erroneously large activation energy. Let us reconsider the central problem of Section 6.2. Again nucleation of a solid from a liquid phase is the best place to start. We shall examine the influence of the vessel which contains the liquid and any foreign matter which may exist therein. Foreign matter is almost certain to be present owing to impurities which, in pure or compound form, are stable above the transformation temperature. If the surface of the container or the impurity is "wet" by both solid and liquid (Figure 3.3), we have the situation depicted in Figure 6.4. The force equilibrium where the three surfaces meet is (Figure 3.3)

$$\gamma_{\text{na}\cdot\text{S}} = \gamma_{\text{na}\cdot\text{L}} - \gamma_{\text{S}\cdot\text{L}}\cos\theta \qquad (6.9)$$

The nucleus will form on the surface of the nucleating agent because a smaller amount of surface energy is needed. Exact quantitative proof of this last statement will not be presented here as much drudgery and little illumination are involved. However, two features are apparent in Figure 6.4 and equation 6.9

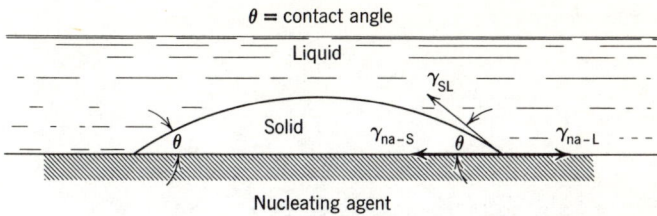

Figure 6.4    Heterogeneous nucleation of a solid from a liquid.

respectively: (a) the na-S surface *replaces* an equivalent amount of na-L surface; (b) $\gamma_{\text{na-S}}$ is less than $\gamma_{\text{na-L}}$.  The appearance of the na-S surface therefore *lowers* the net energy associated with the area it occupies.  The net energy decrease may be subtracted from the S-L surface energy, giving the total surface energy of *heterogeneous nucleation,* which is usually much less than the surface energy for *homogeneous* nucleation despite the nonspherical shape of the nucleus.  By using expressions analogous to those of equations 6.1 and 6.2 we can say, in general,

$$\Delta f^* \text{ (heterogeneous)} < \Delta f^* \text{ (homogeneous)} \qquad (6.10)$$

The rate curve for heterogeneous nucleation must therefore resemble Figure 6.3, except that nucleation begins at much higher temperatures.  Heterogeneous nucleation reduces, and sometimes removes, supercooling.  We may therefore conclude with some assurance that the vast majority of phase changes we observe are nucleated heterogeneously.  In solids, the case for heterogeneous nucleation is still stronger.  As we have pointed out, solid-solid nucleation is accompanied by strains and distortions which increase the activation energy for nucleation substantially.  However, phase boundaries, grain boundaries, dislocations, and similar defects all have their own strain effects, which may be canceled in part by nucleation of a new phase.  In addition (as Chapter 5 indicates), diffusion occurs more readily along such defects than through the crystal lattice, and therefore $\Delta f_D^*$ is reduced.  Consequently, nucleation in solids is invariably heterogeneous.  Figure 6.5 is a typical case: the new phase appears first at the grain boundaries.  Nucleation within the grains may occur on the dislocation substructure.

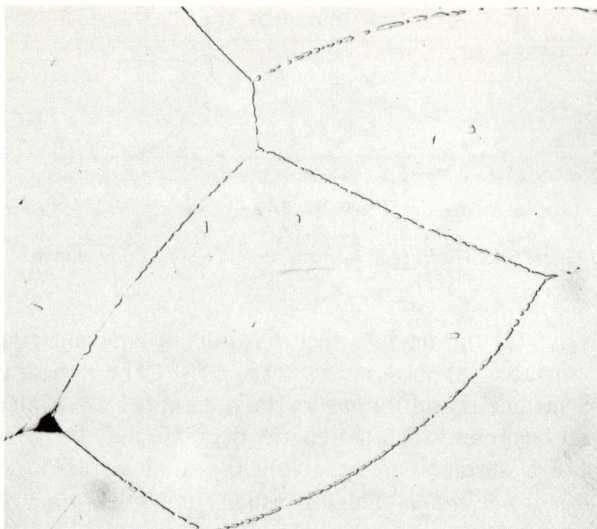

Figure 6.5    Appearance of the new phase at the grain boundaries.

## 6.4    GROWTH OF THE NEW PHASE AND PHASE CHANGE KINETICS

After a nucleus appears it may reduce its total free energy by continuous growth as shown in Figure 6.1. Material is transferred by diffusion in several steps: through the old phase, across the phase boundary, and into the nucleus. The rate of transfer will therefore obey the Arrhenius Equation (equation 4.1), with the activation energy determined by the rate-limiting step in the transfer process. In one-component systems the only step is the phase boundary crossing, and the rate of transfer does not change with time. The boundary therefore moves outward at a constant rate. The volume of a spherical region of new material would be

$$V_n(t) = \tfrac{4}{3}\pi v^3 (t - t_0)^3 \tag{6.11}$$

where $v$ is the boundary velocity, and $t_0$ is the instant of time when the nucleus formed. Nucleation, of course, goes on at all times during the transformation. In the time $dt_0$, $N\, dt_0$ nuclei are formed. The total volume which has transformed is obtained by summing up the volumes of all the nuclei which have formed since the beginning of the transformation ($t_0 = 0$):

$$V = \tfrac{4}{3}\pi v^3 \int_0^t N(t - t_0)^3 \, dt_0 \tag{6.12}$$

However, one more important point should be discussed. Nucleation will not occur in regions which are already transformed. Growth will cease when two phase boundaries meet, completing the transformation at that particular spot. The volume $V$ of equation 6.12 increases indefinitely with time, but we know that it cannot exceed the volume of the entire piece of material which is transforming. If nucleation and growth occur randomly, and a fraction $dy$ of the *untransformed* material is transformed,

$$(1 - x) \, dy = dx$$

where $x$ is the total fraction transformed; $x$ may vary from zero to unity (the end of the transformation). Integrating the above equation and using the boundary conditions on $x$,

$$x = 1 - e^{-y} \tag{6.13}$$

The "uncorrected" volume fraction $y$ may be obtained by dividing the right side of equation 6.12 by the total volume $V_0$. Consequently,

$$x = 1 - \exp\left[-4\pi v^3 \int_0^t \left(\frac{\dot{N}}{V_0}\right)(t - t_0)^3 \, dt_0\right] \tag{6.14}$$

If $(\dot{N}/V_0)$ (nuclei formed per unit volume per unit time) is known, the integration may be completed. Generally, the volume fraction transformed is given by

$$x = 1 - e^{-at^n} \tag{6.15}$$

in which $3 \leq n \leq 6$ normally.

Transformations in multicomponent systems are more complicated. Material must now be transferred through the phases, and bulk diffusion with time-dependent concentrations may control the rate. Consequently, the rate of boundary motion is time dependent, and equation 6.11 must be changed. However, equation 6.15 is usually obeyed with somewhat smaller values of $n$. There are certain general features of all transformations which are worthy of

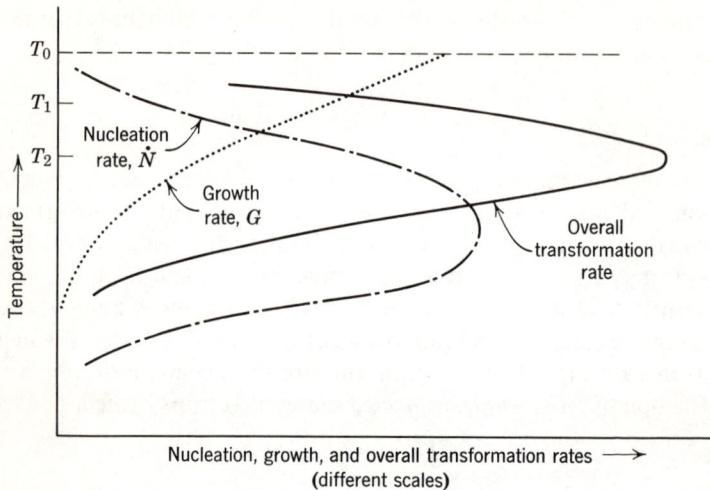

Nucleation, growth, and overall transformation rates ⟶
(different scales)

Figure 6.6   Overall transformation rate.

note, especially their variation with temperature. Figure 6.6 shows the nucleation rate $\dot{N}$ and the growth rate $G$ as functions of temperature. $\dot{N}$ comes from Figure 6.3; $G$ is plotted according to the Arrhenius Equation (equation 4.1) as diffusion will determine the growth rate. The overall transformation rate is proportional to some product of $\dot{N}$ and $G$, and the peak therefore lies at a higher temperature than that of $\dot{N}$. At temperatures near $T_0$ few nuclei are formed but growth is rapid, and the new microstructure will contain few large grains. At the lowest temperatures, the nucleation rate is relatively large, but the growth rate very small. Consequently, low-temperature transformation yields many small grains (i.e., a fine-grained structure).

DEFINITIONS

*Activation Energy for Nucleation.*   The critical free energy plus the activation energy for the diffusion of new material into the nucleus: $\Delta f^* + \Delta f_D$.

*Critical Free Energy ($\Delta f^*$).*   The free energy of a particle having the critical radius, and the maximum free energy of any particle of the new phase.  Subcritical particles lower free energy by redissolving; nuclei, by growing.

*Critical Radius (r\*).*  The minimum size which a particle of the new phase must have in order to grow, that is, to be a *nucleus.*

*Growth.*  Increase of the nucleus in size.

*Heterogeneous Nucleation.*  Nucleation occurring at surfaces, imperfections, severely deformed regions or other structural features which lower the critical free energy.

*Homogeneous Nucleation.*  Nucleation occurring in perfectly homogeneous material such as a pure liquid.

*Nucleation.*  The beginning of a phase transformation, marked by the appearance of tiny regions (called *nuclei*) of the new phase which grow until the transformation is complete.

## BIBLIOGRAPHY

Cottrell, A. H., *Theoretical Structural Metallurgy,* St. Martin's, New York, 1957. Chapter 14 discusses phase change and kinetics on an elementary level.

Kingery, W. D., *Introduction to Ceramics,* Wiley, New York, 1960.  Chapter 10 is replete with examples, data, and photographs of phase changes in ceramic materials.

La Mer, V. K., "Nucleation in Phase Transitions," *Industrial and Engineering Chemistry,* vol. 44 (1952), p. 1270.  A lucid explanation of nucleation kinetics recommended for both graduate and undergraduate.

Turnbull, D., "Phase Changes," *Solid State Physics,* vol. 3, Academic Press, New York, 1956, p. 225.  Exhaustive and comprehensive review article, covering the subject in entirety.  An excellent source of the important references and articles.

## Problems

6.1   Many liquids may be supercooled; that is, under special conditions freezing may be avoided at temperatures well below the equilibrium freezing point.  Explain such behavior and state the conditions and procedures which favor supercooling.

6.2   Derive equations 6.2a and 6.2b and the expressions for $\Delta f^*$ and $r^*$ which were used in Figure 6.2.  Suppose nonspherical nuclei are formed. How would your answer be affected?

6.3   Explain where and why nonspherical nuclei may form.

6.4   Pound and La Mer [*J. Amer. Chem. Soc.* vol. 47 (1952), p. 2323] determined the rate of nucleation of solid tin from its liquid at various temperatures below the melting point by rapidly supercooling a large number ($10^{10}$) of small, tin droplets, separated by an oxide film, and measuring the change in volume of the sample as a function of time. Using the values of free energy transformation per unit volume $\Delta F_v$, from measurements in bulk transformations, they were able to show that the theory of homogeneous nucleation agreed well with their work and that assumptions of a spherical nucleus, and the applicability of the bulk value of the surface energy $\gamma_{SL}$ to nucleus formation, were correct within their accuracy of measurement. Calculate the following quantities from their values of nucleation rate at a constant volume fraction solidified and at 113°C:

(a) The liquid-solid surface energy for tin $\gamma_{SL}$ in ergs/cm$^2$.
(b) The critical radius for the nucleation of solid tin.
(c) The number of tin atoms in a nucleus of critical size.

You are given the following information from this paper:

(1) Taking into account the change in transformation free energy $\Delta F_v$ with temperature, the slope of ln (nucleation rate) versus $1/T$ curve at 113°C was found to be

$$\text{Slope} = -23.8 \times 10^3 \ (°\text{K})$$

for any constant fraction of liquid transformed.

(2) Boltzmann's constant $°k = 1.38 \times 10^{-16}$ erg/°K.
(3) $\Delta F_v = -10^9$ ergs/cm$^3$ at 113°C.
(4) The radius of a tin atom is $1.508 \times 10^{-8}$ cm

6.5   (a) Calculate the critical radius of a BCC $\beta$ phase nucleus in a superheated HCP $\alpha$ matrix at 1175°K in titanium. Assume that the nucleus is spherical and that strain effects may be neglected. (At 1175°K the volume free energy, $\Delta F_v = -6.54 \times 10^7$ ergs/cm$^3$, and the interfacial free energy $\gamma_{\alpha\beta} = 200$ ergs/cm$^2$.)

(b) What is the meaning of a critical radius such as that calculated in part (a)?

(c) If a $\beta$ particle strains the $\alpha$ matrix, would the critical radius be larger or smaller than that calculated in part (a)? Why?

6.6   Graphite may be converted to synthetic diamonds by application of very high pressure and temperature. Lower temperatures may be used if small amounts of nickel are present. Explain. It has been found that the relative lattice parameter of the nickel and graphite impurity is very important. Why? What other effects may be important?

6.7   Do phase changes ever go to completion (i.e., 100% of the old phase transformed)? Explain your answer.

6.8   In countries and eras where beer is quaffed from glassware, it is usually necessary to carbonate the beer in order to provide sufficient "sparkle." Such measures are not necessary if the beer is to be consumed in earthenware mugs; then the naturally dissolved carbon dioxide is sufficient. Explain.

6.9   Would you expect superheating to be as common as supercooling? Explain.

6.10   (a) Explain how the composition of a cast iron melt must be modified to produce nodular rather than flake graphite.

(b) Distinguish between "inoculation" and "nucleation" in making nodular iron.

6.11   How can the grain size of castings be controlled?

6.12   (a) What is meant by epitaxial growth?

(b) Cite some well-known examples.

6.13   (a) Cr films can be deposited on Mo by lowering the temperature of a Cr-Sn liquid solution in which the Mo is immersed.  Explain.

(b) Surface cracks, if present in tungsten, can readily be delineated by the same technique.  Explain.

CHAPTER SEVEN

# Structural Change

SUMMARY

Almost all common structures are not in equilibrium with their surroundings. Control of structure is achieved, therefore, by controlling its
rate of formation and dissolution. In solidification, control of the rates
of nucleation and growth lead to control of grain size and shape, and of
the homogeneity and integrity of the solid material. Similarly, proper
control of the solidification rate may be used to purify. Recrystallization,
which is the nucleation and growth of strain-free material from a plastically deformed matrix, may be similarly controlled. Precipitation hardening, which involves the achievement of a critical dispersion of the
second phase, may require precise controls to be effective; the second
phase may be metastable and always has a nonequilibrium shape and
size. In the heat treatment of steel, a large number of possible structures
may form on the cooling of austenite. The controlled decomposition called
*tempering*, of the as-cooled structure multiplies the possibilities. All common steel structures are nonequilibrium. In all of the cases discussed the
relative rates of formation of the various structures may be analyzed and
predicted by the principles discussed in Chapters 2 and 6.

## 7.1 INTRODUCTION

As we have seen in the preceding chapters, thermodynamic
stability only measures the possibility that a system will attain a
certain state. Whether that state will be part of our experience
must depend largely on kinetics and the rate of its formation.
Almost all practical materials are unstable with regard to their
surroundings. We control structure by controlling the rate of formation and dissolution of structure. That is, we accelerate the
formation and retard the dissolution of the structures we desire.
In this connection, some practices are immediately obvious, such

112

as the use of artificial heterogeneous nuclei. Clouds in the atmosphere are "seeded" with fine AgI crystals, nucleating ice, which turns to rain; very small amounts of nickel accelerate the transformation of graphite to artificial diamond at high pressure and temperature. In the cloud chamber subatomic particles reveal their presence by nucleation of visible droplets from supersaturated vapor. The less spectacular uses are the more common; fine grain sizes can be attained in polycrystalline materials by addition of heterogeneous nucleants to the liquid prior to solidification. Titanium carbide is used with aluminum, aluminum carbide with magnesium, and tungsten carbide with steel. The crystallization of glass is induced by titanium dioxide which has been dissolved in the liquid. The resulting material, much more heat and thermal-shock resistant than glass, is called *Pyroceram*.

Even less obvious than the uses described in the preceding paragraph are the principles and techniques of heat treatment and solidification of metals, which represent a major area of exploitation of the rate of change of phase in liquids and solids. Metallic materials dominate the discussion which follows because the detail and volume of information regarding metallic transformations is greatest, and because heat treatment and solidification of metals have been so important to our culture. The principles involved may be applied to any material whether ceramic, polymeric, or even cementitious. The details will depend on the structure.

## 7.2    SOLIDIFICATION

In Chapter 6, the fundamental details and mechanisms of solidification were used as an illustration of nucleation and growth. We will now examine the practical consequences of solidification. It is apparent from Figure 6.6 that low temperature solidification will lead to fine grain size as the nucleation rate is relatively high and the growth rate low. Therefore, a large number of slowly growing nuclei result from large supercooling or equivalently rapid solidification rates. Conversely, higher temperatures mean higher growth rate and lower nucleation rate, and therefore coarser grain size. Large ingots, which must cool slowly, require additives to induce increased (heterogeneous) nucleation if fine-grained structure is desired. Other problems are also encountered. The first parts of the ingot to solidify are next to the mold walls which readily conduct away the heat. Next is the top of the ingot; consequently the remaining liquid is surrounded by solid. Most liquids shrink when they solidify (water and bismuth are exceptions), and the shrinkage of the liquid in the center of the ingot

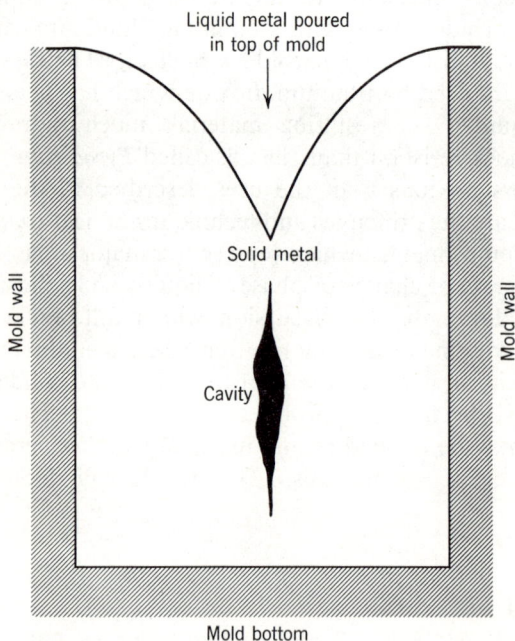

Figure 7.1    Centerline shrinkage in a casting.

leads to shrinkage cavities as shown in Figure 7.1. To avoid such behavior the common practice is to keep the top of the ingot hot, usually by adding compounds which decompose exothermally.

Heat is evolved as the liquid solidifies; the enthalpy of the solid is much less than that of the liquid. This heat must be conducted away through the mold walls. The flow of heat is always perpendicular to the wall surfaces. The first material to solidify, next to the mold wall, is fine grained with grains of random crystallographic orientations. The material in the interior of the ingot cools more slowly, and solidification takes place at a higher temperature. Usually, some of the grains near the surface simply grow inward as heat flows outward. The resulting structure, depicted in Figure 7.2, is termed *columnar*. The columnar grains are not randomly oriented. In general, crystals grow more rapidly in certain directions. Some of the fine grains near the surface

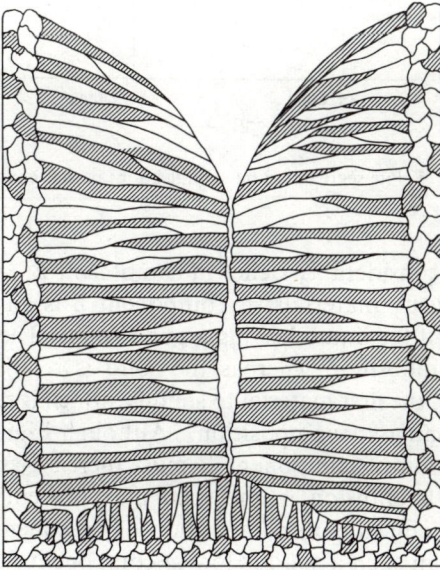

Figure 7.2    Columnar structure in casting shown in Figure 7.1.

have their directions of most rapid growth normal to the mold wall. Heat is withdrawn in this direction, and these grains will grow inward much more rapidly than the others and dominate the solidified structure. Sometimes the columnar grains branch, and the branches branch again. The resulting grains are then referred to as *dendrites*.

Finally, it is important to note that the liquid and the solid are not of the same composition. Consider, for example, Figure 7.3.

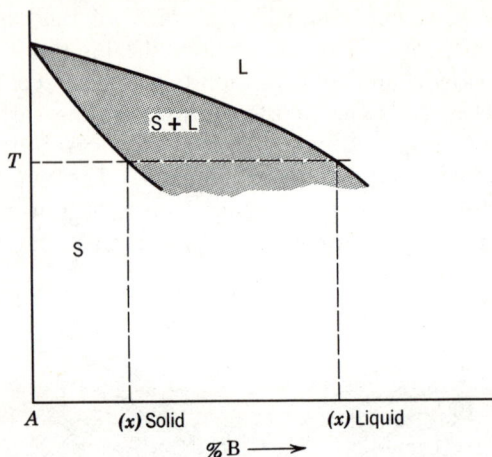

Figure 7.3    Solute segregation on solidification: partial phase diagram.

The liquid is considerably richer in component B.  This fact may be used to purify materials: if component B is an impurity, we simply pour off the liquid leaving a purer solid.  The liquid comes readily to equilibrium, whereas the solid is relatively sluggish. Consequently, the first material to solidify, which is relatively pure, may never change its composition.  Although Figure 7.3 tells us that the fraction of component B must increase as temperature drops, sluggish diffusion in the solid may prevent much of the material from adjusting its composition.  Consequently, the center of each grain in such a solid is low in component B, and the outside of each grain is correspondingly high.  The resulting structure is called *cored.*

Solute segregation of this sort is also useful for purification, in a semi-continuous way.  The first three problems at the end of this chapter deal with the principal methods.  The last of these methods (Problem 7.3) is called *zone refining,* which is capable of purifying materials having suitable phase equilibria to a level of a few parts per billion, and perhaps better.  This fact has made possible the existence of the entire semiconductor industry.  Only at such fantastic purities are transistor materials effective.

Generally, however, nonequilibrium solidification is an annoyance resulting in porous, columnar (or dendritic), cored material, which is usually of very inhomogeneous composition. As we mentioned earlier, fine-grained, noncolumnar structures may be attained by using artificial nucleants. A shrinkage cavity may be eliminated by a "hot top" for the ingot. Coring and gross segregation are usually dealt with after solidification. Heat treatment of the solid at an elevated temperature will homogenize the composition; of course, care is necessary. In the case of Figure 7.3, the last material to solidify has the lowest melting point, much lower than one would expect of the homogeneous solid as it is much richer in component B. Localized melting of the outside of the cored grains may occur with subsequent collapse of the ingot unless the temperature is kept low enough. Under these conditions diffusion will erase composition differences. Sometimes the material is deformed plastically before heat treatment. The deformation itself helps to erase gross composition differences, aids diffusion (by increasing dislocation density), and accelerates recrystallization which we shall discuss next.

## 7.3  RECRYSTALLIZATION

During the plastic deformation of solid materials most of the mechanical work done is expended to change the shape of the particular material and some is dissipated as heat. However, some of the work is stored in the material. This work is actually used to create regions of severe distortion associated with concentrations of dislocations and other defects. To the extent that the energy is stored, the material has departed from equilibrium and will, if possible, return to equilibrium, lowering its free energy by decreasing the number of defects. To accomplish this, in all cases thermal activation is required. There are a number of possibilities. First, any excess point defects (usually vacancies) which may have been generated disappear until the equilibrium concentration is reached. (See Section 5.8 for vacancies in equilibrium.) Vacancies disappear at low temperatures; in copper, for instance, liquid helium temperatures must be used to retain appreciable amounts of vacancies. Higher temperatures may be used with metals of higher melting point (such as tungsten) and some ceramics. Second, at higher temperatures the internal stresses are partially relieved by rearrangement of the dislocation structure. Without detailed discussion we may say that the dislocations move so as to cancel their long-range stress fields. No appreciable decrease occurs in the dislocation density. This second process is called *recovery*. It requires a higher activation energy than point defect annihilation and therefore occurs only at higher temperatures.

Finally, at still higher temperatures the high dislocation density is sharply reduced by the nucleation and growth of completely new, undistorted crystals of the material. This process, called *recrystallization*, completely erases the effects of plastic deformation. (The structural changes which occur have been reviewed in the chapter on microstructure in Volume I.) The rate of recrystallization may be calculated approximately by using the findings of Chapter 6, particularly Sections 6.2 and 6.4.

The decrease in free energy per unit volume $\Delta F_V$ is taken as the difference between the free energies of deformed (and recovered) material and recrystallized material; sometimes $\Delta F_V$ is as high as $10^8$ to $10^9$ ergs/cc. The surface energy is the grain boundary energy, because the nucleus is the same material as the matrix. Equation 6.2 may be applied if $\gamma$ and $\Delta F_V$ are known; $\Delta F_V$ will, of course, depend on the severity of deformation. If deformation is very light, recrystallization may never occur. (Why? See equation 6.2b.) On the other hand, especially severe deformation will increase the nucleation frequency markedly. From equations 6.2b and 6.8, it is clear that increasing severity of deformation, and consequently larger magnitude of $\Delta F_V$, will lead to finer grain size in the final, completely recrystallized material; recrystallization will begin at lower temperatures. The effect of temperature on the recrystallized structure may be predicted from Figure 6.6, and is similar to the effect of temperature on the solidified structures discussed at the beginning of Section 7.2. Lower recrystallization temperatures mean finer grain size and vice versa.

The growth of the recrystallized material and the total transformation are described almost exactly by Section 6.4 and equations 6.14 and 6.15. Typical behavior is shown in Figure 7.4 where $x$ is the fraction recrystallized; the two curves correspond to $n = 3$ and $n = 5$ in equation 6.15. The *incubation periods* ($t_1$ and $t_2$) are

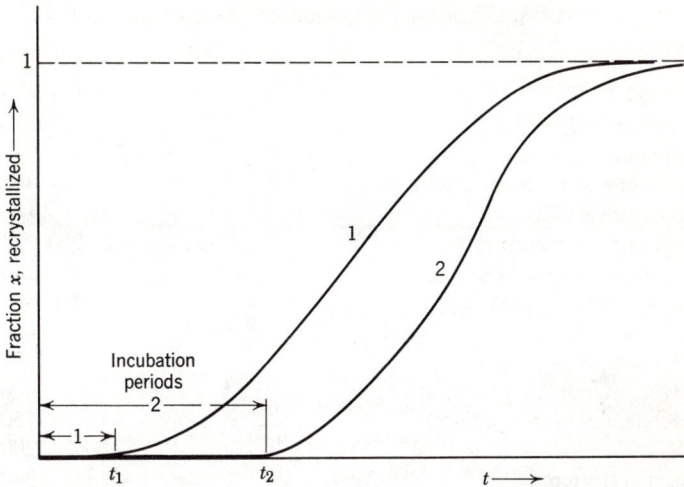

Figure 7.4    Recrystallization versus time for two typical cases.

always observed for all materials. Although such behavior may come directly from equation 6.15, an additional incubation period is observed in the nucleation rate. That is, nucleation is not at first observed. It is possible that the initial nuclei are merely too small to observe. However, recrystallization has one special feature which probably explains the nucleation incubation better; the growth of recrystallized embryos is irreversible. Let us review Section 6.2. Embryos, that is, particles of less than critical size, will "redissolve" for a given phase transformation, and the old phase will reappear. However, the recrystallization embryo cannot redissolve as there is no simple way to recreate the distorted, dislocation-laden structure the embryo has replaced. Consequently, embryos for recrystallization, rather than redissolving, merely wait for another thermal fluctuation to provide them with more energy. Eventually, the critical size is exceeded. The incubation period corresponds to the irreversible growth of the embryos. If the temperature is high enough, nucleation may occur without incubation.

Practically, each material is characterized by a *recrystallization temperature* which is just the minimum temperature necessary to

*Table 7.1    Recrystallization Temperatures (Approximate) for Various Metals*

| | |
|---|---:|
| Lead (99.999%) | Below 0°C |
| Aluminum (99.999%) | 75°C |
| Aluminum (commercial) | 275°C |
| Aluminum (+1% manganese) | 400°C |
| Copper (99.999%) | 100°C |
| Copper (commercial) | 200–250°C |
| Copper (+2% beryllium) | 250°C |
| NaCl (deformed at 400°C) | 450–500°C |
| Iron (pure) | 450°C |
| Molybdenum (sintered) | 1000°C |
| Molybdenum (0.5% titanium) | 1400°C |
| Molybdenum (10% niobium) | 1750°C |
| $CaF_2$ | 1200°C |
| Tungsten (sintered) | 1200°C |

assure complete recrystallization in one hour.  Table 7.1 contains typical data for various ductile materials.  Note particularly that impurities and alloy elements always raise the recrystallization temperature.  All the materials have been deformed below the recrystallization temperature.  Such deformation is called *cold work*.  Deformation above the recrystallization temperature is called *hot work*.  The data in Table 7.1 must be approximate since the materials had differing degrees of cold work prior to recrystallization.  Where high-temperature strength is important recrystallization may be disastrous.  Consequently, a great deal of high-temperature alloy development concerns the raising of the recrystallization temperature.

After recrystallization is complete the grain size may continue to increase.  The larger grains continue to grow and the smaller ones disappear.  In this way the grain boundary area, and consequently free energy, is reduced.  Grain growth is occasionally objectionable.  Sometimes the grain boundaries are needed for diffusion (e.g., for mass transport during sintering, to be discussed in Chapter 8).  In order to obtain strong, dense sintered ceramics, both recrystallization and grain growth must be delayed.  Usually additives are used in a manner roughly analogous to alloying.

## 7.4   PRECIPITATION HARDENING

In a large number of binary systems, especially those of metallic alloys, the solubility of one or both components in the other decreases with decreasing temperature. In essence, the situation is depicted in Figure 7.5. Material of composition $x$ exists as a single solid solution at equilibrium only above $T_0$. If the temperature is suddenly depressed below $T_0$, the solid solution is supersaturated, and may come to equilibrium by the precipitation of $\beta$ phase from the $\alpha$.

The precipitation of $\beta$, which is another case of nucleation and growth, is covered completely, in principle, by Chapter 6. Since $\beta$ is much richer than $\alpha$ in component B, mass transfer of B through the $\alpha$ phase is necessary. As we mentioned in Section 6.4, this feature leads to complications in the quantitative treatment via equation 6.14. However, Figure 6.6 and the qualitative treatment still apply. A coarse precipitate will form at high temperatures and a fine precipitate at lower temperatures. If the temperature is still lower, the nucleation rate is negligible, no transformation occurs, and the $\alpha$ solid solution remains supersaturated. In practice, the control of precipitation has tripled the

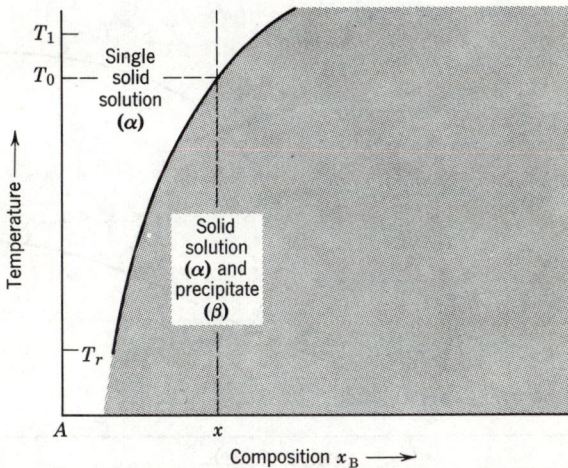

Figure 7.5   Partial phase diagram of a precipitation-hardenable alloy.

useful strengths of aluminum and copper alloys.  Copper-beryllium alloys may be made as strong as some tool steels by proper heat treatment, and *age-hardened* aluminum alloys have been the fundamental, aircraft frame material since the Second World War. These age-hardened aluminum alloys are called *duralumins*.  There are many variations, containing copper, manganese, silver, magnesium, chromium, silicon, iron, zinc in binary, ternary, and higher order mixtures.  In order to obtain high strength, the precipitate must be very fine, as the precipitate particles must interfere with the motion of dislocations.  The dispersion must not be too fine, however, or the stress fields of the particles will not be sufficiently well developed, or on a large enough scale.  A *critical dispersion* will provide the highest strength.

This is the general procedure: the alloy is first heated up into the single phase region (e.g., to $T_1$ in Figure 7.5) until it is homogeneous.  This step is called *solutionizing*.  Quenching follows to temperature $T_r$, which is low enough to prevent precipitation. The material, still relatively soft, is held at $T_r$.  It is hardened by *aging* at some higher temperature.  The time and temperature of aging must be chosen to give the proper precipitate size.  Figure 7.6 describes the general behavior; the hardness is measured as a

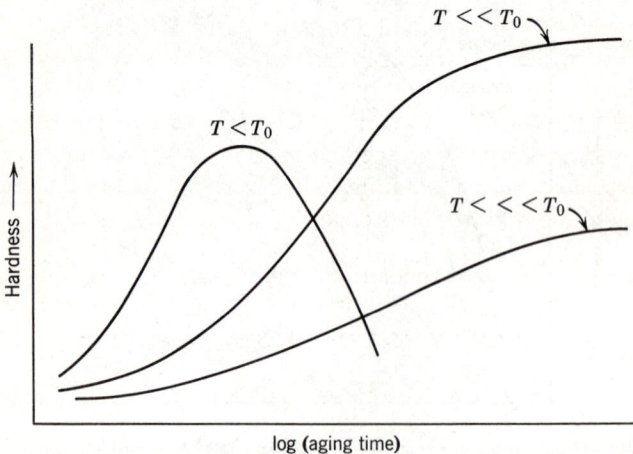

Figure 7.6    Age-hardening: effect of aging temperature.

function of aging time. If temperature is too low ($T \lll T_0$) the critical dispersion is never achieved, although the hardness grows with time as the precipitate size increases. If the aging temperature is too high, the precipitate is too coarse. Also, the initial increase in hardness is wiped out, as in Figure 7.6, if the precipitate size continues to increase. This is called *overaging*. At the appropriate temperature ($T \ll T_0$) maximum hardness is achieved. The material may be quenched or cooled before overaging occurs.

Actually, the first precipitate to form is not the final equilibrium phase; it is merely the structure which forms with greatest rapidity. It is soon replaced by a more stable phase which is not as quick to appear. This stable phase may be replaced in turn by another phase. In this way, the system's free energy is lowered in steps, by passing through a series of *metastable* states on "the road to equilibrium." Such a manner of change is generally the fastest way to proceed to equilibrium and has been studied in detail in the age hardening of aluminum-copper alloys. First, copper atoms congregate, in small clusters less than 100 atoms wide, on the {100} planes of the aluminum-rich terminal solid solution; the clusters are called *Guinier-Preston zones,* or GP-[1]. The clusters grow somewhat and, more important, assume an ordered structure and are then called GP-[2]. Both types may be observed by the very sophisticated x-ray diffraction techniques pioneered by Guinier and Preston. After GP-[2], the phase $\theta'$ appears, which is a form of the compound $CuAl_2$ whose crystal structure is coherent and continuous with that of the matrix phase. Finally, the equilibrium form of $CuAl_2$ ($\theta$ phase), which is massive, and incoherent with the matrix, arrives. Maximum hardness is associated with maximum GP-[2], which is too small to be visible in the optical microscope.

## 7.5   THE HEAT TREATMENT OF STEEL

Section 7.4 described the way in which a supersaturated aluminum-copper alloy may pass through a series of metastable states, and how those states may be achieved and utilized. The most widely used case of metastability, and a major industrial product, are iron-carbon alloys.

The iron-rich side of the iron-iron carbide phase diagram is shown in Figure 7.7. Iron carbide, $Fe_3C$, commonly called *cementite,* is not stable. However, it is metastable at ordinary temperatures and will not decompose for thousands of years. At higher temperatures cementite may decompose to graphite and $\alpha$ iron (usually called *ferrite*); graphite is the equilibrium form for carbon, in this case. However, cementite nucleates and grows much more rapidly than graphite, especially at lower temperatures, and is the more common phase. Consequently, the usual iron-carbon phase diagram is really a working nonequilibrium metastable diagram, for practical use.

Of primary interest in the iron-carbon diagram is the eutectoid reaction,

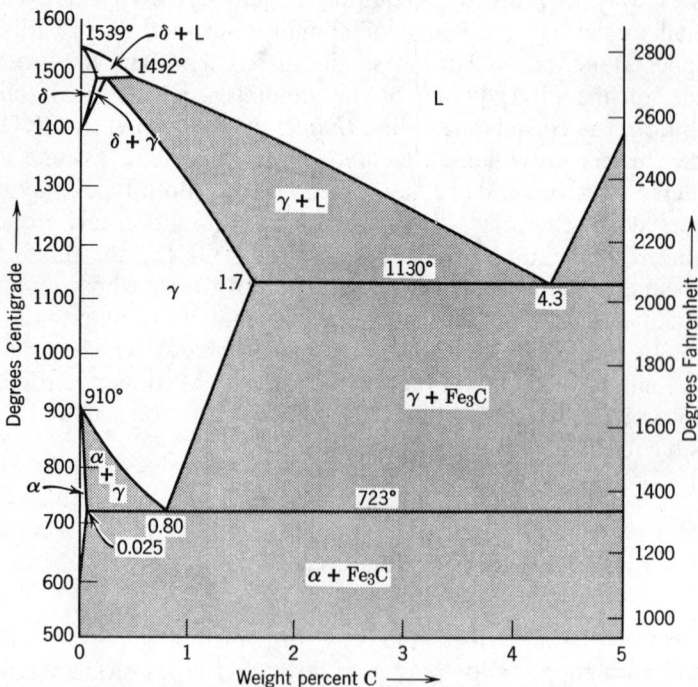

Figure 7.7   The phase diagram for metastable $Fe$-$Fe_3C$.

Figure 7.8    Pearlite.    Magnification, 500×.

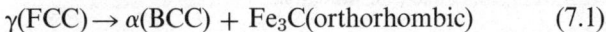

$$\gamma(FCC) \rightarrow \alpha(BCC) + Fe_3C(orthorhombic) \qquad (7.1)$$

which occurs at 723°C.  The temperature is low enough that the rate of formation of graphite is negligible.  The eutectoid structure consists of alternating layers of ferrite and cementite.  Figure 7.8 is a photomicrograph of a steel of the eutectoid composition; the structure is called *pearlite*.  All our previous thoughts on nucleation and growth apply to the formation of pearlite.  Coarse microstructures are formed at high transformation temperatures, and fine pearlite, sometimes beyond optical resolution, at lower temperatures.  The rate of the transformation is described qualitatively by Figure 6.6.  Because the rate is so important, it is determined empirically for each alloy.  Samples are transformed isothermally, and the times at which the transformation of the $\gamma$ phase (called *austenite*) starts and finishes are recorded.  These data are then plotted as a function of temperature.  Essentially, the abscissa is simply the reciprocal of the mean rate of transformation.  The plot therefore resembles an inversion of Figure 6.6, with a minimum transformation time where the rate is a maximum.  Figure 7.9 is such a plot, called a *T-T-T* diagram (time-temperature-transformation), for a steel of eutectoid composition.

Actually, pearlite is formed in this material only above 400 to 500°C.  If the austenite is transformed between 220 and 400°C, the cementite and ferrite grow as extremely fine needles, rather

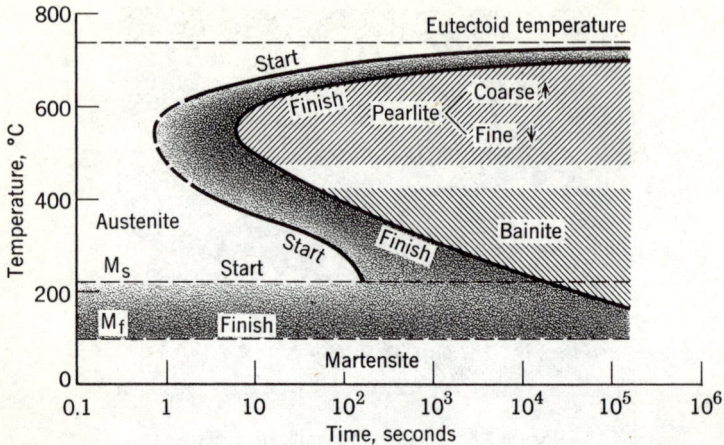

Figure 7.9  A *T-T-T* diagram (time-temperature-transformation) for 0.8% carbon steel.  For isothermal transformation.

than layers.  The needle-like structure is called *bainite*.  The ferrite phase in bainite is usually highly strained.  The strains may stem from the volume changes on transformation and from excess carbon trapped interstitially in the ferrite because of the low transformation temperature.

If the austenite is transformed below 220°C, diffusion and consequently the formation of austenite-cementite mixtures, occur at very slow rates.  A new type of transformation is observed.  The austenite transforms very rapidly to a new, highly distorted phase, called *martensite*.  No diffusion is involved in the martensite transformation, but rather a shearing and expansion of the austenite lattice.  The phase boundary may move at a large fraction of the speed of sound.  The dashed line marked "$M_s$" in Figure 7.9 (martensite start) marks the temperature at which martensite first appears.  Similarly, the line marked "$M_f$" denotes completion of the transformation.  The two lines are horizontal because the transformation is, practically speaking, instantaneous.  The martensite crystal structure is body-centered tetragonal (BCT), which may be viewed as the ferrite lattice (BCC), grossly distorted by carbon atoms "trapped" in solution due to the low diffusion rate.

Although the treatment is rather complex, the martensite transformation, which occurs in a number of alloy systems, may be analyzed within the framework of Chapter 6.

Summing up, austenite may decompose into martensite (lowest temperatures), bainite (intermediate), or pearlite (highest) depending on the transformation temperature. Martensite is extremely hard and is, therefore, a very desirable structure for steels used in tools and machinery of all kinds. In order to obtain martensite and not pearlite or bainite, the austenite must be transformed at a sufficiently low temperature. For instance, the 1080 (0.8% carbon) steel of Figure 7.9 can remain in the 400 to 600°C range no longer than a second or two before beginning to transform to pearlite and bainite. Consequently such a steel must be cooled rapidly, usually by quenching the austenite from above the eutectoid temperature (Figure 7.7) into oil or ice water. The "dangerous" region, which is the dashed portion of the curve in Figure 7.9, is the region of maximum transformation rate; it is sometimes called the *pearlite nose*. However, quenching the hot material in such a violent manner is obviously possible only for small objects. In addition, fast cooling leaves high residual stresses, sometimes cracking the surface and interior of the material. In order to form martensite at slower cooling rates the pearlite nose must be moved to the right. That is, the rates of pearlite and bainite formation must be reduced. This reduction is the purpose of the various alloy elements which are used in "tool steels." A desirable alloy addition is one which moves the pearlite nose to the right without appreciably depressing the $M_s$ and $M_f$. Martensite may then be obtained with lower cooling rates. The ease with which martensite may be formed (in terms of the rapidity of cooling) is called the *hardenability*. Manganese, chromium, nickel, vanadium, and molybdenum, among others, all increase hardenability when added in small percentages; there are *T-T-T* diagrams for all the standard compositions. (See Bibliography at the end of this chapter.)

Once formed, martensite is usually much too brittle to be widely useful. It is therefore heated and allowed to decompose, partially or wholly, into more stable, less brittle structures. This process, called *tempering,* has sequential features similar to the series of age-hardening precipitates of Section 7.4. First, the original untempered martensite breaks up into a special type of low-carbon martensite and a metastable carbide precipitate, called *epsilon carbide,* whose structure is hexagonal close-packed. On further heating the structure then changes to finely divided ferrite and cementite. Finally, after long times or high tempering temperatures, the cementite particles grow and become spherical and fewer in number. The resulting structure is called *spheroidite.* Unlike the age-hardenable alloys, tempered martensite is softer than the untempered material. Generally, tempering represents a compromise between decreased hardness and increased ductility.

### DEFINITIONS

*Age Hardening.*    Precipitation hardening; however, the precipitate is not visible by optical means.

*Austenite* ($\gamma$).    The face-centered cubic form of iron, stable at intermediate temperatures.

*Bainite.*    A structure formed by relatively rapid cooling of austenite, consisting of needles of cementite and highly strained ferrite.

*Cementite.*    The orthorhombic carbide of iron, $Fe_3C$.

*Centerline Shrinkage.*    The cavity developed along the centerline of a casting due to contraction of the metal on solidification.

*Columnar Structure.*    Elongated grains, found in the interior of the ingot, where heat flow is slow and uniaxial during solidification.

*Coring.*    Segregation of components due to nonequilibrium solidification.

*Ferrite* ($\alpha$).    The body-centered cubic form of iron, stable at low temperatures.

*Hardenability.*    The ease of formation of martensite, in a given material.

*Martensite.*    A structure formed by a diffusionless transformation of austenite, when cooling is rapid enough to avoid extensive formation of pearlite or bainite.

*Metastable.*    Unstable but having a very long lifetime.

*Pearlite.*    The lamellar (layered) eutectoid structure of ferrite and cementite, formed from austenite on cooling.

*Precipitation Hardening.*    Strengthening due to nonequilibrium precipitation of a second phase.

*Recovery.*    The relief of internal stress by thermally activated motion of dislocations.

*Recrystallization.*    The nucleation and growth of unstrained regions in plastically deformed material.

*Recrystallization Temperature.*    The temperature at which the material of interest is recrystallized in one hour, or some comparable period.

*Tempering.*    The controlled composition of martensite, by heating.

BIBLIOGRAPHY

*Atlas of Isothermal Transformation Diagrams,* United States Steel Company, Pittsburgh, 1951.

Brick, R. M., and A. Phillips, *Structure and Properties of Alloys,* McGraw-Hill, New York, 1949. A relatively comprehensive volume covering recrystallization, phase diagrams, and microstructural changes in pure metals, copper-base alloys, and magnesium-base alloys. Also covered are heat treatment and hardening of steels, cast irons, tool steels, stainless steels, and many others. Many excellent photomicrographs and detailed data on the alloys themselves.

Chalmers, B., *Physical Metallurgy,* Wiley, New York, 1959. Chapter 6 is probably the most complete and general treatment available of solidification on an undergraduate level. Chapter 8 deals with structural changes in the solid state.

Kingery, W. D., *Introduction to Ceramics,* Wiley, New York, 1960. Part 4, entitled "Development of Microstructure in Ceramics," deals with phase equilibria, phase changes, and metastable structures.

*Supplement to the Atlas of Isothermal Transformation Diagrams,* United States Steel Company, Pittsburgh, 1953. Handbooks of all the *T-T-T* diagrams available at the time of publication.

## Problems

7.1   In the A-B alloy system, the liquidus (L − L + α phase boundary) may be approximated by the line

$$T = T_m - \frac{C_l}{C_E}(T_m - T_E)$$

and the solidus (α − L + α phase boundary) by the line

$$T_s = T_m - \frac{C_s}{C_\alpha}(T_m - T_E)$$

where $C_E$ and $T_E$ are the eutectic composition and temperature, and $C_\alpha$ is the composition of α (A-rich) at the eutectic temperature. $C_s$ and $C_l$ are the compositions (percent B) of the solid and liquid respectively. (This description is merely a quantitative approximation of the left side of a eutectic phase diagram.

(*a*) For $0 < C < C_E$ and $T_E < T < T_m$, find the ratio of $C_s$ to $C_l$. This is called the segregation coefficient *K*.

(*b*) Suppose purification is attempted (i.e., removal of B from the alloy) as follows: the material is heated to $T_E < T < T_m$, and the B-rich liquid is removed. This procedure is then repeated. Find the purity of the solid as a function of the number of repetitions of this procedure *n* and the initial impurity concentration, %B = $C_0$. Is this procedure practical? Why or why not?

7.2   Consider another attempt to purify the B-contaminated A of the previous problem. The material is melted in a cylindrical crucible of length $L$ and cross-section $S$. Then solidification is induced in such a way that a plane liquid-solid interface, perpendicular to the cylinder axis, proceeds down the length of the rod. Assume that mixing is complete in the liquid and therefore that liquid is always homogeneous; assume that diffusion in the solid is negligible. Prove that the concentration of B in the cylinder $C$ as a function of distance along the axis $X$ will be:

$$C = KC_0\left(\frac{L - X}{L}\right)^{K-1}$$

where $K - 1 < 0$. (*Hint:* Conservation of B is the key. Since the interface moves, B is rejected from the solid to the liquid, because $C_s = KC_l$, and so $C_s < C_l$. Therefore, $C_l$ continually rises.) Notice that, at $X = L$, $C$ "blows up." Why? Why does this not occur in a eutectic system? How could this procedure be used to purify materials? Is this practical?

7.3   Consider a third attempt to purify the alloy of the two previous problems. Using the same cylindrical crucible as in the previous problem, let a molten zone of length $Z$ propagate down the cylinder. The two liquid-solid interfaces are perpendicular to the axis. At the "front" interface, solid of composition $C$ is forming. Make the same assumptions regarding mixing in the solid and the liquid as in the previous problem. Assume also that both interfaces move at the same rate, so that $Z$ is constant. Prove that the final impurity distribution will be

$$C = C_0[1 - (1 - K)e^{-KX/Z}]$$

(*Hint:* Again, conservation of B gives the key; the change in total B in the zone should be equal to the difference between that gained by melting new material and that lost by solidification of an equal amount of material of different concentration.)

7.4   The two previous problems outline alternative methods of refining materials. Suppose each operation were performed once on identical bars of alloy. Determine the more efficient method, in total impurity elimination, if the impure half (i.e., the last half to freeze) of each is cut off and thrown away. Assume that $K = \frac{1}{2}$ (for both of course), and $L = 10Z$. Why is the procedure of Problem 7.3, which is called *zone refining*, commonly used to obtain extremely pure materials?

7.5   Can pearlite be formed directly from bainite by heating the bainite? Explain your answer. Can pearlite be formed by tempering martensite? Explain.

7.6    Use nucleation and growth concepts to explain the shape of the *T-T-T* curve.

7.7    If pearlite were annealed for a very long period, just below the eutectoid temperature, what changes would occur?

7.8    The "solutionizing" anneal is used to homogenize age-hardenable aluminum alloys.  Why?  Solutionizing is always carried out below the eutectic temperature, although the phase diagram indicates much higher temperatures are possible before melting.  Why?

7.9    Severely cold-worked copper has a stored energy of $10^9$ ergs/cm$^3$. The grain-boundary energy, $\gamma_{GB}$, for copper is 500 ergs/cm$^2$.  Calculate the critical size and critical free energy for a nucleus of recrystallized material.

7.10    The activation energy for self-diffusion of copper is 47,000 cal/mole.  Referring to equation 6.8, calculate the activation energy for nucleation using the results of Problem 7.9

7.11    Suppose the copper of Problem 7.9 were not so severely cold-worked, so that the stored energy was only $10^8$ ergs/cm$^3$.  What is the activation energy for recrystallization now?  If the copper of Problem 7.9 recrystallizes at 250°C, what is the recrystallization temperature of the copper of this problem?

7.12    The nitrides and oxides of aluminum are very stable.  Steel ingots to which aluminum has been added ("killed ingots") tend to have large centerline shrinkage and very fine grains.  Explain.

7.13    A steel having coarse-grained austenite is more hardenable than one having fine-grain.  Why?

7.14    (a) What phase is maintained by quenching 18:8 stainless steel from 900°C in air?

(b) What phases are present after cold working such a quenched alloy?

7.15    (a) What is retained austenite?
(b) Why is it objectionable?
(c) How is it removed in 18:4:1 high speed steel?

7.16    (a) Describe the Jominy end quench test for steel.
(b) What is its purpose?
(c) What is the McQuaid-Ehn test?

CHAPTER EIGHT

# Sintering

SUMMARY

The bonding together of fine powders or fibers into more or less dense solid bodies is called *sintering*. Usually sintering is accompanied by increased conductivity, mechanical strength, ductility, and, in many cases, density. Often sintering is the simplest and cheapest method of fabrication. Sometimes it is the only method possible. Powder techniques are most commonly employed in ceramics, are used extensively with certain metals, and are coming into use for polymers. The mechanism involved in sintering is the transport of mass by viscous flow, evaporation and condensation, or diffusion by a variety of paths.

## 8.1 INTRODUCTION

Sintering is one of the earliest methods of fabricating metals and ceramics. It is commonly used today to make solid bodies from metal powder, ceramic powder, and, more recently, from some forms of polymer particles. The process of bonding and densifying an aggregate of powder is used when it represents a more economical method of making a particular body or, alternatively, when it is the *only* way in which a certain material can be fabricated into a useful form.

Among those substances for which powder sintering offers almost the only satisfactory manufacturing method are *composite materials:* single bodies which are combinations of several materials or types of materials. Sintered carbide tools, dispersion-hardened materials, fiberglass, ceramic-metal heat resistant materials, metal-graphite friction materials, and polymer or metal impregnated fibers are examples of such composites. The principles underlying

132

the preparation and properties of these composites comprise an entire field of materials science.  Figure 8.1 shows a number of examples of such composites with comments on their properties.

High-melting-point metals such as tungsten, tantalum, molybdenum, niobium, and most ceramics are among the materials which, until recently, could be prepared only by sintering.  Of recent interest is the preparation by powder techniques of porous tungsten to make ion-propulsion engine cathodes and transpiration-cooled surfaces for use in space engines.

There are many variables which can influence the behavior of particles during sintering.  Some, or all of them, may be effective in a particular case and generalities concerning them are not always possible; many questions about the sintering process remain unanswered.  In this chapter, the emphasis will be principally on phenomena that are fairly common.

## 8.2   PROPERTIES OF FINE POWDERS

The adaptability of a given material to fabrication by sintering depends upon the properties of its powder, upon its mode of manufacture, and, of course, upon whether or not the material can be made in the form of a powder.  The properties of powders which are important in this connection include size and size distribution, shape, compressibility, purity, and apparent density.  Many of these properties are related to the way in which the powder was made.

Metal powders are made in various ways.  Frequently a chemical reduction step can be employed in the original refining process which produces a powder directly.  Thus it is often possible to eliminate the melting process entirely in making some metal parts.  This is clearly very advantageous in working with metals whose melting points are high.  Typical reactions which produce metal powders are hydrogen or carbon reduction of oxides, precipitation from solutions or vapors, and electrolysis of solutions.  Both metallic and nonmetallic powders can be made by mechanical fracture and grinding if the material is brittle enough.  Liquid disintegration, or *atomizing,* is a possible method if high enough temperatures and nonreactive crucibles are available.  Sinterable

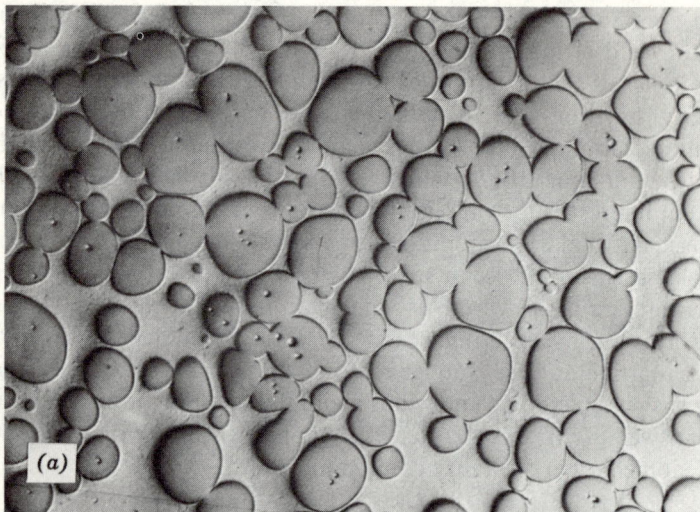

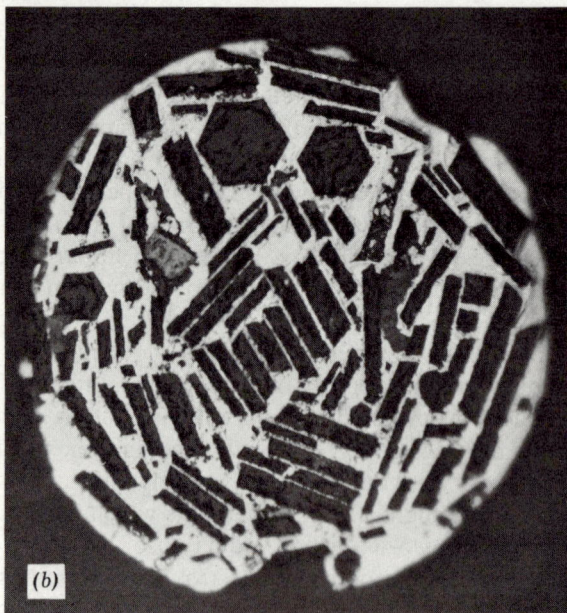

Figure 8.1(a)  A ductile Fe-Ni-W composite.  Spherical particles are W.  (Courtesy L. A. Shepard.)  Magnification, 250×.  (b) A composite of $Al_2O_3$ fibers in a silver matrix.  (Courtesy W. H. Sutton.)

polymers can occasionally be obtained by emulsion polymerization. Regrinding previously fabricated polymers does not produce particles whose surfaces are in the proper condition for sintering.

The size of the particles normally employed for sintering ranges from 0.5 micron to about 200 microns. Materials in such a finely divided form are often highly reactive chemically. Iron, for example, burns readily when freshly prepared in submicron sizes and exposed to air. This spontaneous ignition, known as *pyrophoricity,* is a serious obstacle in preparing some metal powders. It arises from the fact that the ratio of surface area to weight is very large. For instance, one pound of iron particles of one micron diameter has a volume of about six cubic inches, but a surface area of $6 \times 10^4$ square inches. The oxidation of such a large amount of surface evolves heat which, since it is concentrated in a relatively small mass of solid, quickly causes the temperature to rise to the ignition point.

A mass of particles which are small in size, but large in specific surface area, adsorbs a large volume of gases and other impurities. The inclusion or evolution of these impurities has great practical importance in the process of sintering.

The distribution of particle sizes also affects sintering behavior and the properties of the sintered part. The packing behavior of a mixture of particle sizes is shown schematically in Figure 8.2. The volume of a mass of powder of uniform particle size is usually about half occupied by the actual particles and half occupied by the spaces between the particles. When smaller-sized particles are added, they tend first to fill these spaces and then to increase the overall volume. A powder sample of mixed particle size is generally more dense than one of a single size. Spherical particle shape and a narrow range of particle size are desirable when the end product is to be porous (e.g., a filter).

## 8.3    PREPARATION FOR SINTERING

Preparing a powder for sintering normally involves pressing or consolidating it in order to bring adjacent particles into intimate contact and to break through surface films. The method depends on powder shape and compressibility, which in turn are related to

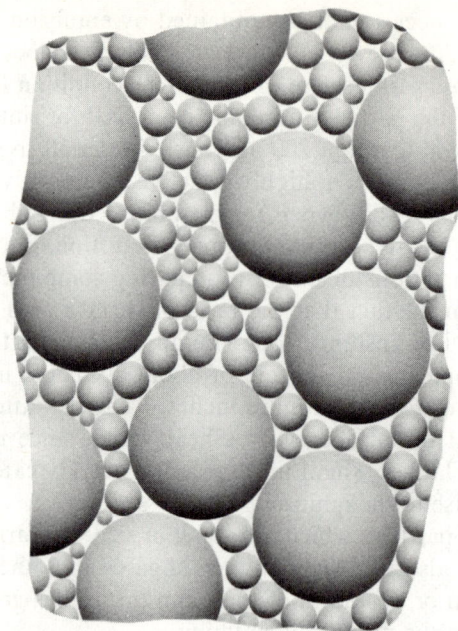

Figure 8.2    Schematic representation of packing of a mixture of particle sizes.

the way in which the powder was made, and to the strength of the material itself.  Pressing techniques which may be used include cold pressing in dies, extrusion with binders, *isostatic* pressing in a flexible envelope under a high pressure in a fluid, and *slip casting,* in which a slurry of powder is dried in an appropriate shape by an absorbent die material.  During normal powder pressing operations many variables, such as the pressure level, friction caused by the die wall, ejection stresses, and the work-hardening characteristics of the powder itself, must be taken into account.  Proper die design and use of binders are arts in themselves and differ greatly with different materials.

It is desirable in some cases to sinter loose powders directly; for instance, this is one way to produce filters with a maximum porosity.  Another situation where loose powder sintering is useful is in the fabrication of beryllium, which because its crystal structure is anisotropic, becomes oriented in an undesirable manner when fabricated in usual ways. When loose beryllium powder is sintered directly, the original random crystal orientations of the powder particles are preserved.

## 8.4 RESULTS OF THE SINTERING OPERATION

Sintering is effected when powder particles are brought together at a temperature sufficient to bond them together. For some materials, such as silver-mercury amalgam (dental fillings) and ice, temperatures at or below room temperature are high enough. Usually, however, a higher temperature must be used, and a wide variety of heating techniques are employed to achieve it. Frequently a controlled or protective atmosphere is necessary.

The sintering of a material usually causes many changes in its properties. In ceramics strength, thermal conductivity, density, transparency, and translucency are increased. In polymers density and strength are increased. In metals conductivity, strength, and ductility usually increase; whether density increases or decreases depends upon the details of the preparation of the metal. A decrease in density usually is the result of trapped gases evolved from the powder surface or of the decomposition of lubricants trapped in closed pores. In compacts containing two components, expansion can result also from diffusional porosity (Section 5.7). The property changes during sintering are the result of modification of size, shape, and amount of porosity as the temperature is brought to the point at which mass transport occurs. The temperature involved is usually near the recrystallization temperature of the base material. It would seem, therefore, that sintering mechanisms involve atom movements similar in scale to those of recrystallization. Generally, the average distance of atomic migration is roughly that of the particle size.

## 8.5 MECHANISMS OF SINTERING—SINGLE-PHASE MATERIALS

Though the ratio of surface area to volume of a particle depends to a certain extent on its shape, in general it varies inversely with the diameter of the particle. The total free energy of a large number of fine particles is therefore greater than that of a smaller number of coarse particles or a solid block of equal solid volume. It is thus reasonable to assume that the driving force for fine-powder sintering is the reduction of total surface area and total free energy.

An understanding of the rate at which various particles sinter is facilitated by making calculations and experiments on the kinetics of bonding between spheres of dimensions larger than typical powders, and then relating these to a powder compact. A sintering sphere model which has proved to be very useful experimentally is indicated by the two cases shown in Figure 8.3. From the spherical geometry of the particles in this figure the neck radius $x$ can be related to the particle radius $r$ in case ($a$), and the decrease in center-to-center distance between particles $2h$ is case ($b$). From these data the volume of material $V$ which was moved in order to form the neck and the area of interparticle contact $A$ can be computed. These are tabulated in Table 8.1 for the two cases shown in the figure. From these calculations the degree of sintering of the two spheres can be characterized by the experimental measurement of neck radius and decrease in center-to-center distance.

The rate at which the neck volume changes is an indication of the rate of sintering and is determined by the rate at which atoms move into the neck region. Possible mechanisms of mass transport are indicated schematically in Figure 8.3, which is really a comprehensive review of known means of atom movement. Some processes which give rise to the movement shown in case ($a$) are (1) evaporation from a convex pore surface and condensation on the concave neck surface, and (2) surface diffusion between the same regions on the solid-vapor surface. Processes producing case ($b$) are (1) viscous or plastic flow of a solid, (2) volume diffusion from the region of interparticle contact into the neck, and (3) grain boundary diffusion between the same regions along the interparticle boundary. The rate at which each of these processes transports a given volume of material may be approximated by a variety of flux equations. Equating this mass transport flux to the geometric requirement governing sintering spheres, summarized in Table 8.1, gives a series of relations between neck radius $x$ and

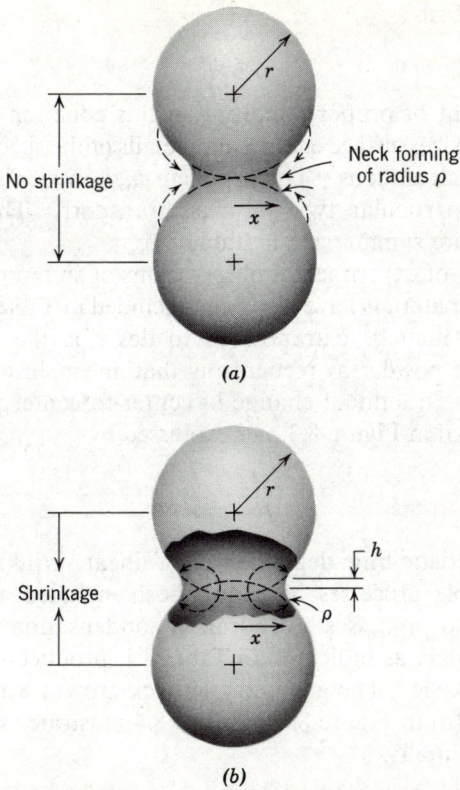

Figure 8.3    Mass transport paths in the sintering of spheres.

Table 8.1    *Geometric Relations in Sintering Spheres*

|  | $V$ | $h$ | $A$ | $\rho$ |
|---|---|---|---|---|
| Case (a) (no shrinkage) | $\dfrac{\pi x^4}{2r}$ | 0 | $\dfrac{\pi^2 x^3}{r}$ | $\dfrac{x^2}{2r}$ |
| Case (b) (shrinkage) | $\dfrac{\pi x^4}{4r}$ | $\dfrac{x^2}{2r}$ | $\dfrac{\pi^2 x^3}{2r}$ | $\dfrac{x^2}{4r}$ |

time, of the form

$$x = ct^m \tag{8.1}$$

The coefficient of proportionality $c$ in this equation depends on temperature, size, surface energy, and details of the specific assumptions in the calculations. The exponent $m$, however, is characteristic of the particular type of mass transport. These various calculations are summarized in Table 8.2.

The results of experimental observations of sintering spheres of a number of materials have also been included in Table 8.2. These observations may be extrapolated to describe the sintering of masses of fine powder by recognizing that linear shrinkage $\Delta L/L_0$ is equal to the fractional change in center-to-center particle distance as shown in Figure 8.3 and expressed by

$$\frac{\Delta L}{L_0} = \frac{h}{r} \tag{8.2}$$

The characteristic time dependences of linear shrinkage for each of the possible processes have also been included in Table 8.2. There are two processes, evaporation-condensation and surface diffusion, which, as indicated in Table 8.1, produce neck growth but no shrinkage. The geometry of neck growth was characterized by case (a) in Figure 8.3. Figure 8.4 illustrates such a situation experimentally.

*Table 8.2   Sintering Relations*

| MECHANISM | $m$ | $n$ | OBSERVED MATERIAL |
|---|---|---|---|
| Viscous flow | ½ | 1 | glass |
| Evaporation-condensation | ⅓ | $\Delta L/L_0 = 0$ | NaCl |
| Grain boundary diffusion controlled by vacancy formation | ¼ | ½ | — |
| Volume diffusion | ⅕ | ⅖ | Cu, Ag |
| Grain boundary diffusion controlled by vacancy movement | ⅙ | ⅓ | — |
| Surface diffusion | ⅐ | $\Delta L/L_0 = 0$ | ice |

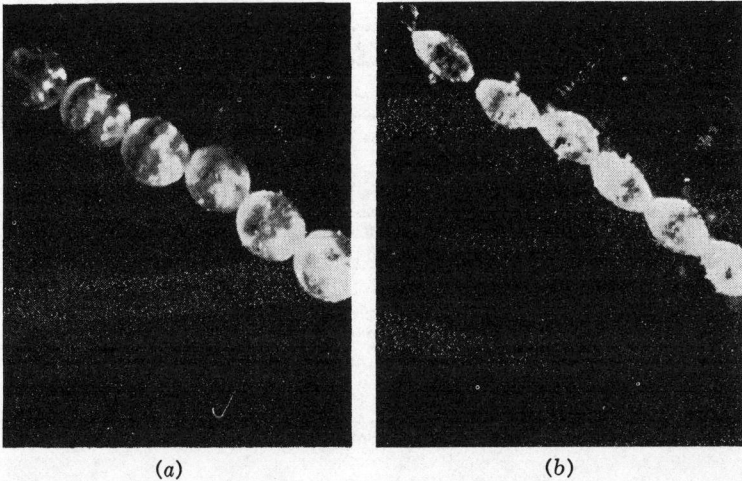

(a)                                             (b)

Figure 8.4    Photomicrographs of sintering sodium chloride at 750°C: (a) 1 min., (b) 90 min. Figure 12.17 From *Introduction to Ceramics* by W. D. Kingery (Wiley, 1960), p. 374.

## 8.6   MECHANISMS OF SINTERING—MULTIPHASE MATERIALS

In Section 8.5 we discussed the sintering of a single-phase material in which any existing pores were filled by the vapor of the material itself. Usually, however, the situation is not so simple as this. The presence of a second phase can accelerate an alternative mass transport mechanism which would have been of secondary importance in the pure material.

The presence of a liquid phase capable of dissolving some of the solid particles produces a transport path geometrically the same as that of the grain boundary path in solid-phase sintering. This equivalence has been observed in copper-iron, tungsten-carbide-cobalt, and a number of ceramic oxide systems. Figure 8.5 represents, schematically, the progress of linear shrinkage during liquid-phase sintering. The first and initially rapid shrinkage occurs at the time the liquid component melts, and its amount depends on the volume of liquid which forms. The second stage of shrinkage occurs when the atoms of the solid component diffuse

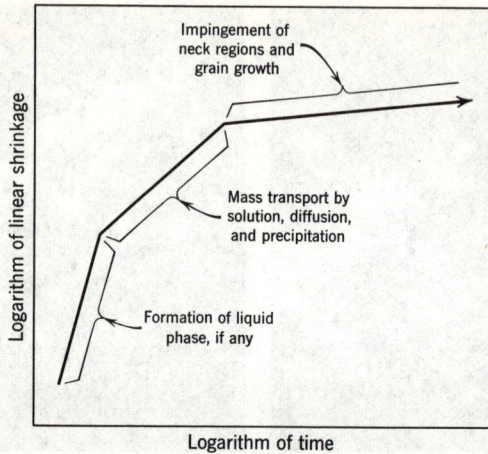

Figure 8.5    Linear shrinkage during liquid-phase sintering.

through the interparticle liquid film, a path which is analogous to the grain boundary path in solid-phase sintering. The final and slower shrinkage occurs when the numerous interparticle contacts on a single particle begin to compete with one another for diffusing material.

The competition of particle contacts causes a decrease in sintering rate in both solid-phase and liquid-phase sintering. When this point has been reached it is easier to visualize the structure as pores on grain boundaries rather than as isolated contacting particles. In this late stage of sintering it has been established that the presence of grain boundaries near pores permits the densification of the material to continue. The inhibition of grain growth by the deliberate use of foreign additives, for instance, permits complete densification of $Al_2O_3$. Materials other than liquids can accelerate one of the possible mass transport paths. Additions of volatile halides to iron or small amounts of water vapor to tungsten or molybdenum increase bonding rate by transport of metal atoms through the vapor phase. The addition of solid metals such as nickel or palladium, in amounts less than one percent of the total weight, accelerates the grain boundary sintering mechanism in tungsten and molybdenum. Sintering may be carried out successfully at temperatures hundreds of degrees lower than those necessary for the pure metals.

## 8.7  EXPERIMENTAL OBSERVATIONS

The experimental measurement of neck growth and shrinkage during sintering can be correlated to calculations made using the mechanisms postulated in Section 8.6. However, the real validity of such an analysis must be established by simultaneous, experimental observations of other features of the sintering process. The role of grain boundaries as a source of material diffusing into pores in metals and $Al_2O_3$ has also been mentioned in Section 8.6. This phenomenon could theoretically be caused, however, either by a grain boundary transport mechanism or a volume diffusion mechanism. The question has been settled by the simultaneous observation of porosity forming inside the particles of the faster-diffusing species in binary metallic systems. Since this porosity could only be caused by diffusion (the Kirkendall Effect), the operation of the volume diffusion mechanism in these cases has been confirmed. This behavior has recently been recognized as one of the reasons that copper-nickel and copper-iron mixed powders *expand* during sintering. Similar additional evidence of volume diffusion has been found in the sintering of several metals and oxides. Diffusion coefficients computed from sintering data appear to agree well with those determined by other means.

### DEFINITIONS

*Composite Materials.*  Solid bodies made of at least two dissimilar materials.

*Isostatic Pressing.*  Pressing powder under a gas or liquid so that pressure is transmitted equally in all directions.

*Pyrophoricity.*  The tendency of a fine particle to react with an environment which would not react with an equivalent amount of material in a completely solid form.

*Sintering.*  The process by which fine particles of a material become bonded together. The process is generally accompanied by an increase in strength, ductility, and occasionally by an increase in density.

## BIBLIOGRAPHY

Jones, W. D., *Fundamental Principles of Powder Metallurgy*, St. Martin's, New York, 1960.

Kingery, W. D. (Editor), *Ceramic Fabrication Processes*, M.I.T. Technology Press and Wiley, 1958.

Kingery, W. D. (Editor), *Kinetics of High Temperature Processes*, M.I.T. Technology Press and Wiley, 1959.

Leszynski, W. (Editor), *Powder Metallurgy*, A.I.M.E. and M.P.I., Interscience Publishers, New York, 1960, 1961.

## Problems

8.1   Why are some fine metallic powders pyrophoric?

8.2   Spheres of equal size occupy 74% of total volume when closely packed, leaving 26% void.  A typical loosely packed collection of such spheres contains 40% void.  If two volumes of spherical powder, each of uniform particle size but one particle size much larger than the other are mixed, what overall volume fraction of each powder produces the maximum mixed density?  Consider two cases: close packing and typical loose packing.

8.3   Explain why spherically shaped and flake-shaped particles often exhibit very low as-pressed strength (green strength) while angular and irregularly shaped particles, when pressed, are stronger.

8.4   How may spherical models be used to examine sintering mechanisms?

8.5   What numerical relationship exists between the geometry of sintering spheres and a sintering mass of powder?

8.6   What evidence exists besides measurements on sph~~ ~ to indicate that metals sinter by volume diffusion?

8.7   By what means can sintering mechanisms be deliberately altered?

8.8   (a) Sketch the apparatus of Hall, which is used to atomize aluminum powder.

(b) Sketch a process for the atomization of iron alloys by water. Where does the greatest wear occur?

8.9   (a) Plot an alloy diagram between WC and Co or TiC and Ni.

(b) Indicate schematically the following: (1) presintering; (2) final sintering of carbide tools made from the above.

8.10   (a) Describe how polymer-bonded diamond cut-off wheels are made.

(*b*) Describe how bronze-bonded diamond cut-off wheels are made.

8.11   (*a*) Describe an electrolytic process for making iron particles of less than domain size.

(*b*) Describe a mechanical method for making iron filaments of less than domain thickness.

8.12   (*a*) How can whiskers of $Al_2O_3$ be made in quantity?

(*b*) How can they be bonded with metal?

8.13   (*a*) Hot co-extrusion of powder mixes in evacuated metal envelopes is a useful technique in various fields both for densification and for hot fabrication.   Discuss the economic feasibility of the process for manufacture of Cr wire.   Include data on the kind of powder you would use; how you would encapsule it, to what size you would extrude it and how you would draw it down to a fine size.

8.14   (*a*) Discuss different methods for making 18:8 stainless steel powder.

(*b*) Why should it be difficult to press cold?

(*c*) At what temperature does it sinter readily in (1) purified hydrogen; (2) tank hydrogen?

8.15   Transpiration-cooled tubes can be made of 18:8 steel powder or from 18:8 wire utilizing certain manufacturing techniques.

(*a*) Describe how you could control pore size in the powder product and in the wire product.

(*b*) What do you imagine the relative strengths and ductility of both products are?

CHAPTER NINE

# Oxidation

SUMMARY

The rate of oxidation depends on temperature and on the properties of the material and its oxide. Even if oxidation causes large decreases in free energy, the oxide layer may be sufficiently protective to retard oxidation. If the volume of the oxide formed is nearly that of the metal consumed, and the oxide is solid, adherent, nonreactive, and nonvolatile, the oxide will probably be protective. Oxygen ions, metal ions, or electrons must diffuse through the protective oxide layer in order to carry on the reaction. The oxide thickness varies logarithmically, parabolically, or linearly with time, depending on the degree of protection afforded by the oxide. Certain phases or regions in metals or alloys may oxidize selectively, leaving the bulk of the material untouched.

## 9.1 INTRODUCTION

The practical use of many materials, particularly at elevated temperatures, is frequently limited by their detrimental reaction with their environment. The most familiar example is the reaction of metals with air. There are many other reactions, however, which do not actually involve oxygen but which resemble the metal-oxygen reaction in their details. Among these are reactions between components in liquid-metal-cooled heat exchangers, rocket nozzles in contact with combustion products, materials for handling chemicals, and ceramics in contact with other materials. Because of this kinetic resemblance to the metal-oxygen reaction, these reactions are normally included in the general subject of oxidation.

The only type of environmental attack of this nature which is not covered in this chapter is the aqueous corrosion of metals.

146

This reaction occurs by quite different mechanisms from the metal-oxygen reaction in the absence of water, although both involve an increase of positive valence and are therefore called oxidation reactions.    Aqueous corrosion will be considered separately in Chapter 10.

## 9.2   THE DRIVING FORCE FOR OXIDATION

The tendency of a metal to react with oxygen is indicated by the free energy change accompanying the formation of its oxide. Table 9.1 lists a number of these values for some common metals. Oxidation can occur if it is accompanied by a free energy decrease. Conversely, if the free energy of oxide formation is positive, the metal will not oxidize.    Most of the metals in Table 9.1 show a negative free energy of oxide formation and will therefore react with oxygen.    Indeed, most of them occur as oxides or other compounds in nature, and most of them oxidize more or less readily when exposed to air.

As with any spontaneous process, oxidation is thermodynamically possible if it decreases free energy; the rate at which it occurs, however, is governed by kinetic factors.

Table 9.1    Free Energy of Formation of Metal Oxides (Per Oxygen Atom) at 227° C (500° K) in Kilocalories

| Calcium | − 138.2 |
|---|---|
| Magnesium | − 130.8 |
| Aluminum | − 120.7 |
| Titanium | − 101.2 |
| Sodium | − 83.0 |
| Chromium | − 81.6 |
| Zinc | − 71.3 |
| Hydrogen | − 58.3 |
| Iron | − 55.5 |
| Cobalt | − 47.9 |
| Nickel | − 46.1 |
| Copper | − 31.5 |
| Silver | + 0.6 |
| Gold | + 10.5 |

## 9.3    SURFACE FILMS

Studies in electronics, catalysis, and corrosion have demonstrated the extreme difficulty of obtaining a perfectly clean metallic surface. In some cases treatment in ultrahigh vacuum, accompanied by heating or ion bombardment, can produce a clean surface. (By ultrahigh vacuum, we mean pressures below $10^{-10}$ mm. of mercury.) However, if such a surface is exposed even to moderately high vacua, it quickly becomes filmed with an invisible layer of the gas. If the gas is chemically inert, it forms a loosely bound layer on the surface. This layer, whose formation is due to secondary van der Waals attractive forces between the surface metal atoms and the gas, is said to be physically *adsorbed*. Gold, silver, and the platinum metals become coated with an adsorbed oxygen layer when exposed to air. When most other clean metals are in contact with oxygen, however, a chemical compound is formed on the surface. This may be a true stoichiometric compound one or more molecules thick, or it may be a monatomic oxygen layer which is said to be *chemisorbed*. Additional amounts of gas may be physically adsorbed on such a layer. Oxide films will in general continue to thicken, particularly at high temperatures. Exposure to ultrahigh vacuum will readily remove a van der Waals adsorbed gas layer, but higher temperatures or ion bombardment are necessary to remove an oxide film.

## 9.4    MECHANISMS OF FILM GROWTH

An initial film of oxide forms very rapidly on clean metal surfaces exposed to oxygen. Two general types of behavior are observed among those metals whose oxides are stable. The alkali and alkaline earth metals (sodium, potassium, magnesium, etc.) form a porous oxide film. Other metals, such as iron, copper, and nickel, form more dense films. The thickness and the growth rate of these films determine the degree to which the oxide protects the metal beneath it.

Whether the oxide formed is dense or porous can be correlated to the ratio of oxide volume produced to metal volume consumed in oxidation.  This ratio, known as the *Pilling-Bedworth Ratio,* is formulated:

$$\frac{\text{Volume of oxide}}{\text{Volume of metal}} = \frac{Md}{amD} \qquad (9.1)$$

in which $M$ is the molecular weight of the oxide having the formula $Me_aO_b$ and density $D$, and $m$ is the atomic weight of the metal which has density $d$. The number of atoms of metal per molecule of oxide is given by $a$. It should be noted that the values of the two densities alone are not sufficient to express the relative volume relationship.

If the ratio of volumes in equation 9.1 is less than one, a volume of oxide forms which is smaller than the volume of metal oxidized. For this reason the oxide tends to be porous and nonprotective. If the ratio is near, or greater than one, the oxide will not be porous. Provided that additional criteria discussed in Section 9.5 are satisfied, an oxide-metal volume ratio near unity indicates that an oxide is likely to be protective, while a ratio much greater than unity indicates that compressive stresses will be produced and the oxide may buckle and crack off, a process called *spalling*.  Figure 9.1 shows a violent case of spalling.

Figure 9.1   Spalling during the elevated temperature oxidation of a prospective high temperature alloy.   The Pilling–Bedworth ratio is much greater than unity. (Courtesy N. J. Grant.)

The mechanism by which the oxide film thickens has been studied by Wagner and his co-workers.  They have delineated several typical cases which differ in the rates and paths by which the atoms of oxygen and metal come together to react. These are illustrated in Figure 9.2.

(1) If the oxide film first produced is porous, molecular oxygen can pass into its pores and react at the metal-oxide interface. This situation prevails when the volume ratio is less than one, as in alkali metals. Figure 9.3 shows the consequences of a porous oxide film.

(2) If the oxide film is not porous, the oxidation reaction may occur at the air-oxide interface. In this case, the metal ions diffuse from the metal-oxide surface outward to the air-oxide interface. Electrons also migrate in the same direction to complete the reaction.

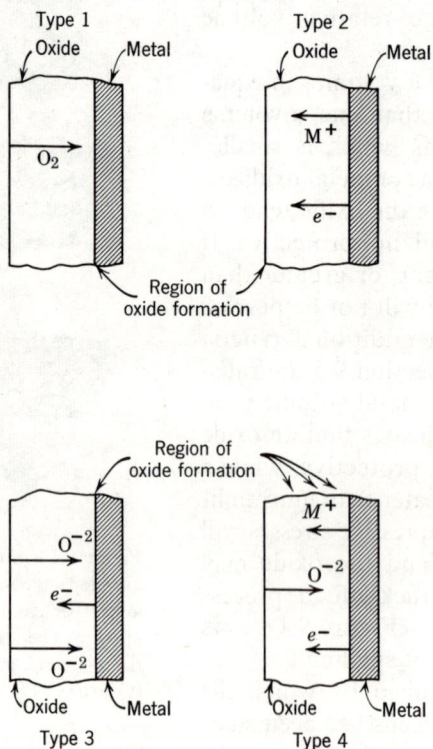

Figure 9.2   Mechanisms of oxidation.

Figure 9.3   Progressive stages of elevated temperature oxidation in an alloy whose oxide is porous. (Courtesy N. J. Grant.)

(3) The reaction may also occur at the metal-oxide interface when the film is nonporous.   In this case oxygen ions diffuse into the film to react at the metal-oxide interface and electrons must, again, be free to move to the air-oxide interface.

4) The final possible mechanism is a combination of cases (2) and (3) in which oxygen ions diffuse inward and both metal ions and electrons move outward.   Here the reaction site may be anywhere in the oxide film.

Each of the last three cases requires that ionic diffusivity and electrical conductivity in the oxide be high.   If its electrical conductivity is low, its rate of growth is limited by the small number of electrons moving from the metal to the gas-oxide interface. The growth of the film may also be slowed down if the rate of passage of the positive metallic ions is slow compared to that of the electrons, as this causes the formation of a *space charge* of positive ions in the film which opposes further diffusion of positive ions into the film.   Since both the diffusivity and the electrical conductivity of oxides are often dependent upon the composition of the oxide, different concentrations in the metal can lead to different oxidation rates.

## 9.5   RATE OF FILM GROWTH

As oxidation proceeds, the oxide layer increases in thickness and the thickness of the metal decreases.   For a few limiting cases we may easily calculate the increase in oxide depth (or decrease in metal depth) with time, for oxidation of a plane surface. If the oxide is nonprotective, direct attack occurs at the metal surface, at a more or less constant rate, and therefore if $x$ is the oxide thickness,

$$\frac{dx}{dt} = \text{(constant) and}$$

$$x = \text{(constant)}t + \text{(another constant)} \qquad (9.2)$$

If the oxide film is protective, diffusion of some kind, metal or oxygen ions or electrons, is necessary to transfer material across the film. The last three cases in Figure 9.2 are typical of parabolic oxidation mechanisms. The compositions at the faces of the oxide are constant, and we may apply Fick's First Law of diffusion (equation 5.1), as we will in Problem 9.5, to derive the relation,

$$\frac{dx}{dt} = \frac{\text{(constant)}}{x} \text{ and}$$

$$x^2 = \text{(another constant)}t + \text{(still another constant)} \qquad (9.3)$$

If the oxide is very protective, and has relatively low electrical conductivity, a third rate law applies:

$$x = A \log (Bt + C) \qquad (9.4)$$

where $A$, $B$, and $C$ are constants. This behavior is attributed to the buildup of an immobile layer of charge, positive or negative, electrons or ions, which retard further diffusion. Iron and nickel oxide films, following this rate law at low temperatures, do not exceed 100 to 200 angstrom units in thickness, after very long times. Aluminum and beryllium oxides also grow logarithmically. Chromium oxide conducts electrons readily but not ions. Consequently, the metal ions in the oxide build up, and a logarithmic growth results, despite the high conductivity for electrons.

## 9.6   PROTECTIVE OXIDES

In Sections 9.4 and 9.5 the porosity and conductivity of oxide films were shown to influence oxidation rates. These factors, therefore, are also related to the protective nature of an oxide. If an oxide is to be protective, the volume ratio must be near unity and its conductivity must be low.

From a practical standpoint adherence is as important in a protective oxide as its imperviousness. The protective nature of oxide films on aluminum, for instance, is due to the fact that aluminum oxide forms a scale which is coherent with the underlying metal. In fact, some atoms in this scale are part of both the oxide and the metal. Strong adherence is also helpful in cases of large volume ratio, where stress is chiefly responsible for any breakdown of the film. If the unconstrained volume of oxide is greater than that of the metal from which it was formed, the freshly formed oxide film is in a state of considerable lateral compression and may well fail by blistering, shear cracking, or flaking. The shape of the metal surface, the relative coefficients of expansion of the oxide and metal, and the rate of heating and cooling of the metal influence the adherence and thus the protective nature of the film.

To be protective, an oxide must also be nonvolatile and nonreactive with its environment. At elevated temperatures, molybdenum and tungsten form volatile oxides and oxidize catastrophically in air. Chlorine, bromine, iodine, and fluorine also form volatile corrosion products with many metals at relatively low temperatures.

Temperature influences the protective nature of an oxide. The growth rates of many metal oxides which follow the parabolic law are described by the Arrhenius Equation (equation 4.1). Since *parabolic oxidation* implies diffusion, and the coefficient of $t$ in equation 9.3 contains a diffusion constant (shown in Problem 9.5), the activation energy for oxidation may turn out to be that of the diffusion step. In such a case a metal which was serviceable for a long period of time at low temperatures may oxidize extremely rapidly if the temperature is increased. The nature of oxide growth rate itself may change with changing temperature. The oxides of some metals grow according to the logarithmic law at low temperatures, the parabolic law at intermediate temperatures, and the linear law at high temperatures. The temperature ranges, in which the changes in the governing rate law occur, depend both upon the identity and the environment of the base metal. Figure 9.4 shows a series of stainless steel samples which have been heated in air in the presence of various compounds. In some

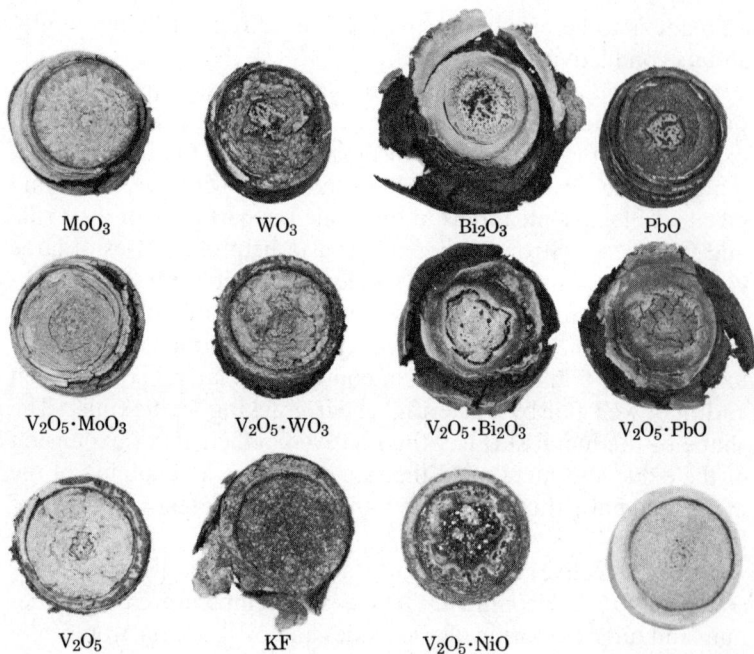

Figure 9.4   Type 347 stainless steel, contaminated with the substances shown, and heated to 1900°F for 24 hours in air.   Specimen at lower right is not contaminated. (Courtesy N. J. Grant.)

cases, attack has been accelerated owing to destruction of the steel's protective film. The transition to the linear rate law at high temperatures is usually associated with cracking in the oxide film.

The degree of protection which an oxide offers its base metal will in all cases depend on temperature and environment and must be observed experimentally. Such observations permit the specification of conditions under which a given metal may be employed and suggest several means by which metals may be protected from oxidation. Surface coatings of other metals or compounds can be employed to separate the base metal from its oxidizing environment. Cracking or flaking of such a coating may be caused by wear, thermal shock, or other means and is the most frequent cause of failure of the part. Other materials may be added to the base metal which alter the relative volumes of oxide and metal, the electrical conductivity and diffusivity of the oxide, or, as in the case of chromium, permit the formation of an oxygen-rich surface film. This technique is useful in designing oxidation-resistant alloys, but in many cases is limited because the amount of alloy required is excessive.

## 9.7    SPECIAL EFFECTS OF OXIDATION

*Selective oxidation* occurs when one component or structural constituent of an alloy oxidizes more readily than the others. This happens in binary solid solution alloys when one component has a much higher negative free energy of oxide formation than the other.

Selective oxidation can be beneficial. The presence of the chromium in many high-temperature alloys insures their high resistance to oxidation. Similarly, the silicon in cast iron improves its oxidation resistance. Copper-base alloys containing aluminum can be *internally oxidized* to produce particles of $Al_2O_3$ in the matrix. The resulting composite is hardened by the dispersed oxide and is exceptionally strong at temperatures near the melting point of the base metal. Selective oxidation, however, may also be harmful. For example, the carbon in carbon steel oxidizes more rapidly than the iron; thus at high temperatures the surface becomes decarburized and is therefore less hard.

In many alloys the grain boundaries are often selectively oxidized long before the grains themselves have been attacked. Such *intergranular oxidation* can thin the cross-section of a metal part and thus cause its mechanical failure at high temperatures.

Nonmetallic materials may also oxidize. Silicon carbide resistance-heating elements may only be operated in air at temperatures at which an $SiO_2$ film can form. If these elements are operated below 1100°C or above 1500°C, the oxidation products are subject to reactions with their environment similar to those to which metal oxides undergo. When two otherwise solid oxides come in contact they may *flux* one another by forming a eutectic liquid of low melting point. Polymers are also subject to destructive attack by oxygen. Rubber, for instance, becomes embrittled by degradation even at low temperatures because of the reaction of the oxygen with unsaturated bonds in the polymer chain. Most polymers will decompose and burn if held at a sufficiently high temperature in the presence of oxygen.

## DEFINITIONS

*Internal Oxidation.*  The selective oxidation of a dispersed phase which is more reactive than the matrix phase.
*Oxidation.*  The increase in positive valence of a metallic element during a chemical reaction; here used to refer principally to reaction with oxygen.
*Pilling-Bedworth Ratio.*  The ratio of the volume of oxide formed to the volume of metal consumed during oxidation.
*Protective Oxide.*  An oxide which forms readily and covers a metal surface to such an extent that it essentially stops further reaction.
*Rate Laws.*  The relationships between oxide thickness and time of oxidation.
*Selective Oxidation.*  The preferential attack by oxygen of one of the components or phases in an alloy.

## BIBLIOGRAPHY

Birchenall, C. E., *Physical Metallurgy,* McGraw-Hill, New York, 1959.
Guy, A. G., *Elements of Physical Metallurgy,* Addison-Wesley, Reading, Mass., 1959.
Kubaschewski, O., and B. Hopkins, *Oxidation of Metals and Alloys,* Butterworths, London, 1953.
Wagner, C., "Diffusion and High-Temperature Oxidation of Metals," *Atom Movements,* American Society for Metals, Cleveland, 1951.

## Problems

9.1  Platinum and chromium are both resistant to detrimental oxidation; explain why.

9.2  According to the Pilling-Bedworth Ratio, which of the following oxides would not be protective?

| OXIDE | OXIDE DENSITY (GM/CC) | ATOMIC WEIGHT OF METAL | METAL DENSITY (GM/CC) |
|---|---|---|---|
| $Na_2O$ | 2.27 | 23 | 0.97 |
| FeO | 5.70 | 55.8 | 7.87 |
| $Fe_3O_4$ | 5.18 | 55.8 | 7.87 |
| $Fe_2O_3$ | 5.24 | 55.8 | 7.87 |
| $Al_2O_3$ | 3.70 | 27.0 | 2.70 |
| $U_3O_8$ | 7.31 | 238.0 | 18.70 |

9.3   What factors other than a favorable oxide-metal volume ratio determine whether a particular oxide protects the base metal against further oxidation?

9.4   What types of measurements can be made to determine the amount of oxidation?   Why is the simple measurement of oxide thickness often not sufficient?

9.5   Show that when diffusion of ions controls the oxidation rate, the parabolic rate law prevails.

9.6   Generally, metals whose oxide growth is parabolic are preferable for use in an oxidizing environment; yet some metals exhibiting linear growth rates are also useful in such environments.   Suggest reasons for these facts.

9.7   Explain the oxidation resistance of stainless steel.

9.8   List and explain ways in which alloying elements influence oxidation resistance.

9.9   Heat-resistant alloys containing chromium are often used in temperature measuring devices.   Frequently these devices are inserted into a furnace chamber in a closed-end ceramic tube so that the alloy itself is isolated from the furnace atmosphere.   Occasionally the alloys oxidize rapidly, forming a green scale.   Suggest reasons for this behavior.

9.10   Stainless steels containing 18% chromium and 8% nickel are susceptible to intergranular attack if slowly cooled from above 1200°C. Explain this behavior, considering the reaction of chromium with other elements in the steel, and the nucleating sites of any reaction products.

9.11   The refractory metals if used in air at 1200°C fail readily as turbine bucket materials.   Show how Nb, Ta, W, and Mo fail in such an application.

9.12   Why could you produce a highly porous sponge of iron whose density is 0.5g per cc starting with mill scale hardwood charcoal and barium carbonate, if you used a temperature of 1000°C and nearly closed containers.

9.13   Account for zinc-coated niobium being rather oxidation resistant at 900°C.

9.14   The ash of Venezuelan fuel oil contains vanadium pentoxide. Why then should the life of 20% chromium alloys be very short when heated with Venezuelan oil at 1000°-1250°C?   How could you prevent such catastrophic corrosion?

9.15   Cobalt-bonded tungsten carbide tools wear rapidly when used to cut steel.   Nickel-bonded titanium carbide tool containing minor amounts of Nb and Mo stand up well.   Explain.

9.16   Why do aluminum brake drums exhibit catastrophic wear?

9.17   What is meant by the friction cutting of steel as compared to cutting with an abrasive wheel?

9.18   Stainless ingots can be "cropped" by cutting with a hot flame containing iron oxide.   Explain.

9.19   A cheap, heat-resistant iron tube can be made by flame spraying an iron tube with stainless steel and painting the same with a mixture of water glass and aluminum pigment.   Explain the nature of the bond and the mechanism of protection of such a coating.

9.20   Describe the following processes: (1) calorizing, (2) siliconizing, (3) chromizing.

CHAPTER TEN

# Aqueous Corrosion

SUMMARY

When dissimilar metals in electrical contact are immersed in an electrolyte, aqueous corrosion occurs. One metal, called the *anode,* dissolves as positive ions. The electrons generated in this way flow through the electrical contact to the other metal (called the *cathode*), where they are used to reduce hydrogen ions, create hydroxyl ions, or participate in a similar reduction reaction. Consequently, the corrosion rate is proportional to, and associated with, an electrical current. Whether a metal is anodic or cathodic depends on its propensity to ionize. It is common to express the propensity to ionize in terms of voltage rather than free energy. A tabulation of such voltages is called an electromotive force (*EMF*) series. The voltages depend on the concentrations of ions in solution, and the shielding of the cathode by reaction products (*polarization*). For any given environment, a table which places the materials in order of their anodic tendencies may be constructed. Such a table is called a *Galvanic Series.* Differences in microstructure, plastic strain, access to oxygen, ion concentration, or anything else affecting the free energy of ionization, will lead to corrosion. Certain materials are passive, that is, they possess extremely stable oxide layers on their surfaces, only in certain environments. Corrosion may be controlled by cathodic polarization, passivation, inhibition, isolation, and cathodic protection.

## 10.1 INTRODUCTION

The aqueous corrosion of metals occurs when metallic atoms dissolve as ions in an aqueous environment. Since this happens when dissimilar metals are placed in electrical contact in the presence of an electrolyte, corrosion is said to be electrochemical in nature. The term "dissimilar" is used to indicate that different free energy changes occur when equivalent quantities of each

159

metal are ionized and dissolved into the environment from the respective metal surfaces. As we shall see, dissimilarity may arise from concentrations of the ions in the environment, the electrolyte velocity, the microstructure of the metals, stress, surface films, and from many other causes. Few fields of materials science are ignored more than corrosion. The yearly economic loss from corrosion is extremely large.

## 10.2   ELECTROCHEMISTRY

Consider the cell shown in Figure 10.1, and the following reaction:

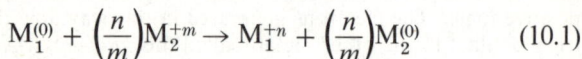

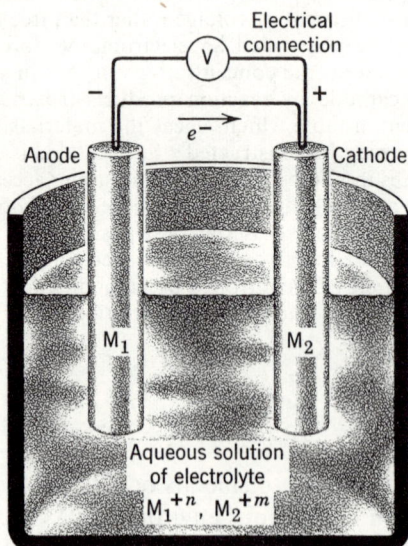

$$M_1^{(0)} + \left(\frac{n}{m}\right)M_2^{+m} \rightarrow M_1^{+n} + \left(\frac{n}{m}\right)M_2^{(0)} \qquad (10.1)$$

Figure 10.1   An electrochemical cell. Negative terminal of voltmeter is connected to the anode.

If the free energy change for the above reaction is negative, $M_1$ (the anode) will be ionized, and the resulting electrons will flow through the electrical connection to participate in the reduction of $M_2$ ions. It is the free energy change which provides the necessary work to transfer the electrons from $M_1$ to $M_2$; that is, a voltage appears between $M_1$ and $M_2$ which depends on the free energy change. If the cell is reversible, the entire free energy change may be utilized. It is obvious that in reaction 10.1, $n$ moles of electrons are transferred for every mole of $M_1$ ionized. It is known that one mole of electrons constitutes about 96,500 coulombs of charge. This number is known as *Faraday's Constant,* which is usually denoted by the letter $\mathscr{F}$. Consequently, the free energy change and the electrical work may be equated, for a reversible cell:

$$\Delta F = -n\mathscr{E}\mathscr{F} \tag{10.2}$$

In equation 10.2 the free energy change is in units of joules per mole, and the voltage $\mathscr{E}$ from $M_2$ to $M_1$ is in volts. The negative sign on the right side is assigned so that the voltage is positive when the free energy change of equation 10.1 is negative. This procedure is conventional in America; in Europe the opposite is practiced.

Equation 10.2 enables us to derive thermodynamic data from the voltages of reversible cells; free energies follow directly. Using equation 1.16, and the Gibbs-Helmholtz Equation (see Problem 1.6), we derive

$$n\mathscr{F}\left[\frac{\partial \mathscr{E}}{\partial T}\right]_{p=\text{const.}} = \Delta S \tag{10.3}$$

As we shall see, it is more convenient to break down the reaction in equation 10.1 into two *half-cell* reactions:

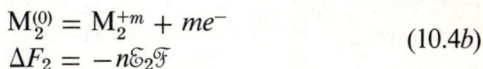

$$\begin{aligned} M_1^{(0)} &= M_1^{+n} + ne^- \\ \Delta F_1 &= -n\mathscr{E}_1\mathscr{F} \end{aligned} \tag{10.4a}$$

$$\begin{aligned} M_2^{(0)} &= M_2^{+m} + me^- \\ \Delta F_2 &= -n\mathscr{E}_2\mathscr{F} \end{aligned} \tag{10.4b}$$

Consequently, the reversible potential of any cell may be calculated from the "half-cell potentials."

## 10.3  THE ELECTROMOTIVE SERIES

According to Section 10.2 the potential between the anode $M_1$ and the cathode $M_2$ in Figure 10.1 may be represented as the sum of two half-cell potentials. The voltage across any complete reversible cell may be obtained by addition of the two appropriate half-cell potentials. Half-cell potentials, having been determined experimentally, are available in tables such as Table 10.1, which is called an *electromotive force* (*EMF*) *series*. The EMF series, however, is determined by complete cell measurements, which are, of course, the only kind we can make. Therefore, the sums (or differences) of the half-cell potentials are defined, but not the magnitudes. In order to define the magnitudes, it is necessary to assign, arbitrarily, a potential to a single half-cell reaction, which then serves as a standard, and determines all the other half-cell voltages. It is customary to assign a zero voltage to the ionization of hydrogen, as in Table 10.1.

*Table 10.1    Electromotive Force Series*

| ELECTRODE REACTION | STANDARD ELECTRODE POTENTIAL $\mathcal{E}$ (VOLTS), 25°C | |
|---|---|---|
| $Na = Na^+ + e^-$ | +2.712 | Active (more anodic) |
| $Mg = Mg^{+2} + 2e^-$ | +2.34 | |
| $AL = Al^{+3} + 3e^-$ | +1.67 | |
| $Zn = Zn^{+2} + 2e^-$ | +0.762 | |
| $Cr = Cr^{+3} + 3e^-$ | +0.71 | |
| $Fe = Fe^{+2} + 2e^-$ | +0.440 | |
| $Ni = Ni^{+2} + 2e^-$ | +0.250 | |
| $Sn = Sn^{+2} + 2e^-$ | +0.136 | |
| $Pb = Pb^{+2} + 2e^-$ | +0.126 | |
| $H_2 = 2H^+ + 2e^-$ | 0.000 | |
| $Cu = Cu^{+2} + 2e^-$ | −0.345 | |
| $Cu = Cu^+ + e^-$ | −0.522 | |
| $Ag = Ag^+ + e^-$ | −0.800 | |
| $Pt = Pt^{+2} + 2e^-$ | −1.2 | |
| $Au = Au^{+3} + 3e^-$ | −1.42 | |
| $Au = Au^+ + e^-$ | −1.68 | Noble (more cathodic) |

## 10.4   CONCENTRATION AND CELL POTENTIAL

Because the free energy per mole of any dissolved species depends on its concentration (see Problem 2.7), the free energy change and electrode potentials of any cell must depend on the composition of the electrolyte. Consequently, the half-cell potentials of Table 10.1 apply only for the standard ion concentration of one molal solution. From the law of mass action (and Problem 2.7), we expect that ion concentrations in excess of the standard will tend to push the reactions of Table 10.1 to the left, thereby decreasing the voltages. The concentration change has altered the free energy change of each reaction. It is therefore possible to construct a cell whose anode and cathode are of the same material, as in Figure 10.2. The electrode on the left will tend to dissolve because of the lower ion concentration, and therefore becomes the anode.

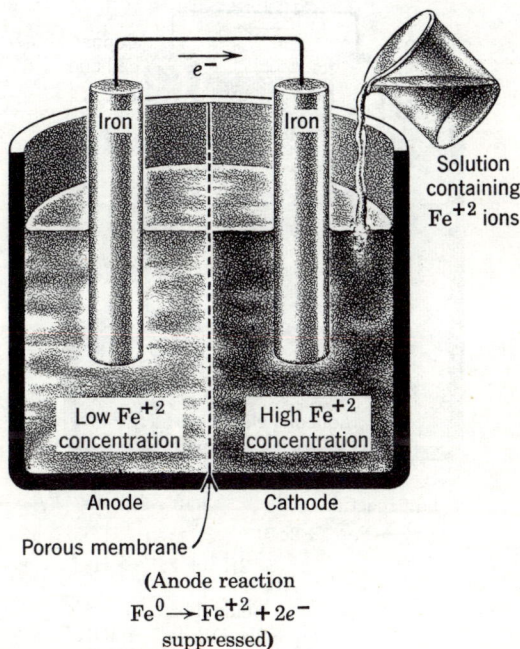

Figure 10.2   An ion concentration cell.

## 10.5    THE CATHODE REACTION, POLARIZATION, AND SURFACE FILMS

Suppose the cell of Figure 10.1 were changed, so that very few $M_2$ ions were present in the electrolyte. Reduction of these ions would therefore no longer be a possible cathode half-cell reaction for this case, but there are other possibilities. Consider the cell of Figure 10.3. The concentration of copper ions is soon reduced to a very low level, as Table 10.1 shows the iron to be "anodic" to the copper. The cathode reaction is then found to depend on the acidity of the solution. In acid solution hydrogen ions are reduced at the cathode. In neutral or basic solution hydroxyl ions are created from dissolved oxygen and water.

In acid solution, the neutral hydrogen atoms created at the cathode will *plate out* on the surface, effectively shielding or *polar-*

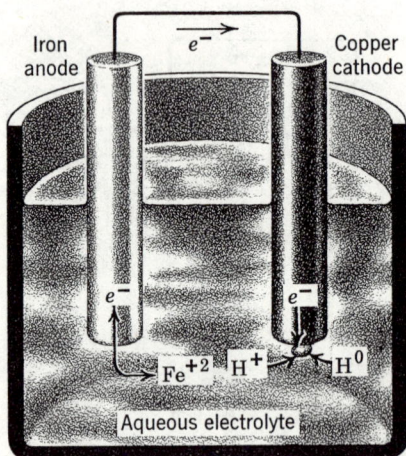

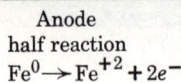

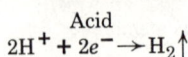

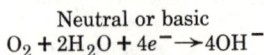

Anode
half reaction
$$Fe^0 \rightarrow Fe^{+2} + 2e^-$$

Cathode
half reactions
Acid
$$2H^+ + 2e^- \rightarrow H_2\uparrow$$

Neutral or basic
$$O_2 + 2H_2O + 4e^- \rightarrow 4OH^-$$

Figure 10.3   Currents and reactions in an iron-copper galvanic cell.

(Cathode reaction
$$O_2 + 2H_2O + 4e^- \rightarrow 4OH^-$$
accelerated)

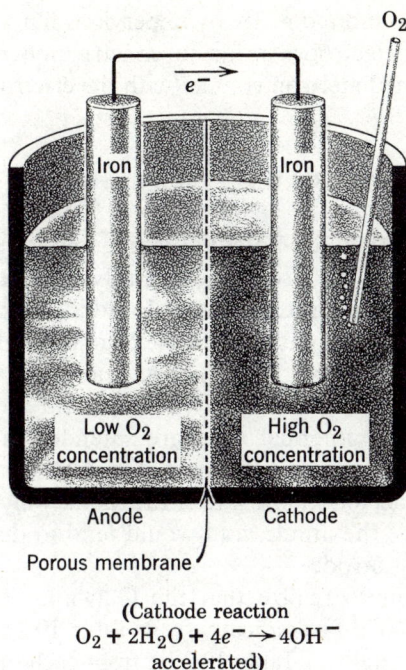

Figure 10.4    An oxygen concentration cell.

*izing* the cathode, and decrease the potential of the cell. However, the atoms usually form molecular hydrogen gas by nucleation, which effectively *depolarizes* the cathode; the properties of the cathode surface are therefore important. The main effect in neutral or alkaline solutions, where the cathode reaction is

$$O_2 + 2H_2O + 4e^- \rightarrow 4OH^- \qquad (10.5)$$

is exhaustion of the oxygen concentration at the cathode surface and an excess of hydroxyl ions. It is the general rule that polarization of either type limits the current flow through the cell. In neutral or alkaline solution, depolarization is accomplished by supplying oxygen to the cathode area. We might therefore construct a second type of concentration cell, as in Figure 10.4. The iron electrode to the right is the cathode due to the high oxygen concentration which expedites reaction 10.5. Sometimes the cell current is limited more directly by impervious films which may be present on the electrodes prior to construction of the cell, or which form immediately on contact with the electrolyte.

## 10.6  THE GALVANIC SERIES

We have seen that cell potential is not only a function of the electrode materials, but also of the electrolyte, because of ion concentration and polarization effects. In addition to the standard EMF series (Table 10.1), it is desirable to obtain cell potentials in a single, common environment such as sea water. Such data, in tabular form, are called a *Galvanic Series* and are presented in Table 10.2. No numerical data are intended, as the table is intended for practical use. If a pair of metals from this list were electrically connected in sea water, the metal which is higher on the list would be the anode, and would tend to dissolve as ions, that is, it would corrode.

It is important to realize that the Galvanic Series for other environments may differ radically from Table 10.2. Also, the farther away two metals on Table 10.2 lie from each other, the larger is the tendency to corrode.

## 10.7  AQUEOUS CORROSION

Recapitulating, the following situations may give rise to electrochemical potential differences which cause the anode to corrode:

(*a*) Difference in chemical composition.
(*b*) Difference in metallic ion concentration, in the electrolyte.
(*c*) Difference in depolarization, due to different concentration in the electrolyte of gaseous oxygen.

We have not, however, exhausted our possibilities.

From Section 10.2, it is easy to see that any chemical change having a negative $\Delta F$ will have an associated electrochemical potential. For instance, one portion of a piece of metal may be plastically deformed, and therefore have a higher free energy per unit volume than the rest of the piece. Consequently, the

*Table 10.2    Galvanic Series in Sea Water*

CORRODED END (ANODIC, OR LEAST NOBLE)

Magnesium
Magnesium Alloys
Zinc
Galvanized Steel or
Galvanized Wrought Iron
Aluminum
(52SH, 4S, 3S, 2S, 53S-T in this order)
Alclad
Cadmium
Aluminum
(A17S-T, 17S-T, 24S-T in this order)
Mild Steel
Wrought Iron
Cast Iron
Ni-Resist
13% Chromium Stainless Steel Type 410 (Active)
50-50 Lead Tin Solder
18-8 Stainless Steel Type 304 (Active)
18-8-3 Stainless Steel Type 316 (Active)
Lead
Tin
Muntz Metal
Manganese Bronze
Naval Brass
Nickel (Active)
Inconel (Active)
Yellow Brass
Admiralty Brass
Aluminum Bronze
Red Brass
Copper
Silicon Bronze
Ambrac
70-30 Copper Nickel
Comp. G-Bronze
Comp. M-Bronze
Nickel (Passive)
Inconel (Passive)
Monel
18-8 Stainless Steel Type 304 (Passive)
18-8-3 Stainless Steel Type 316 (Passive)

PROTECTED END (CATHODIC, OR MOST NOBLE)

deformed region is anodic, and the undeformed remainder is cathodic. Similarly, fine-grained material is anodic to coarse-grained material. Differences in stress level even in the elastic range also lead to potential differences. We therefore list:

(d) Difference in stress, plastic strain, and microstructural imperfections.

Also, most obvious,

(e) Direct chemical attack by the environment.

The results of the situations listed above are all around us. As an example of (a) observe Figure 10.5, the consequence of contact between magnesium and steel.

Differences in fluid velocity may allow metallic ions to build up at points of low velocity. The ion concentration difference (b) makes the high-velocity points anodic. Figure 10.6 is typical of the result. Oxygen concentration difference (c) is even more common. In a partially full metal tank the fluid nearest the "water line" is most strongly oxygenated, and the nearby metal will be cathodic to the rest. Crevices having less access to oxygen are anodic to the rest of the metal and are severely attacked, as in Figure 10.7. As the corrosion products (e.g., "rust") build up at the anode the oxygen supply is further curtailed, accelerating the reaction. Scale, rust, or any other deposit therefore lead to further attack. Barnacles, a familiar marine organism, act similarly. Figure 10.8 is a good example.

Severely cold-worked metals are susceptible to corrosion of type (d), as in Figure 10.9, which exemplifies *stress corrosion*. Notice that the crack follows the grain boundaries, which are anodic to the interior of the grains. Even if the brass had been annealed, the stresses encountered during actual use may lead to failure. Stress concentrations in the brass lead to corrosion at the points of concentration; stress concentration and the process then accelerate catastrophically. It is fortunate that only a limited number of electrolytes lead to stress corrosion. For brass, for instance, ammonia and mercurous salts are the culprits. Grain boundaries, even without the presence of stress, may corrode. We shall discuss a notable case of *intergranular corrosion* in Section 10.8. A second phase of different composition sets up a potential differ-

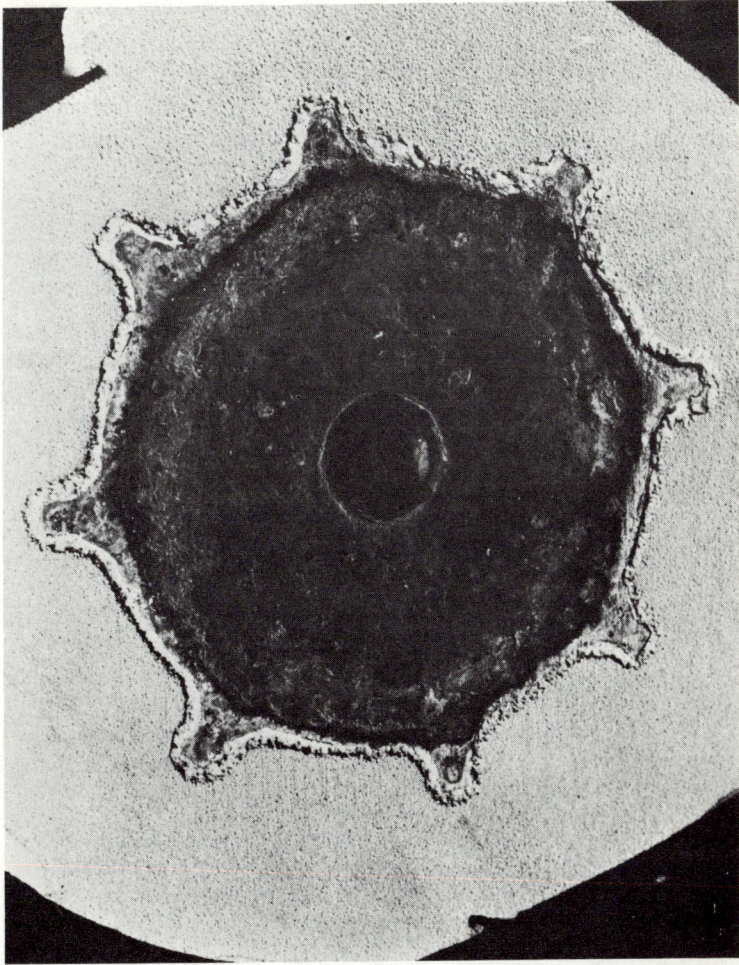

Figure 10.5    Severe aqueous corrosion of magnesium in contact with steel. (Courtesy of F. L. La Que, International Nickel Company.)

ence which may lead to corrosion.  The "age-hardened" aluminum alloys discussed in Chapter 7 are very susceptible to attack by marine atmospheres because of the presence of the second phase. For general use in aircraft the alloy is coated with pure aluminum. The composite is called *Alclad*.

## 10.8    THE CORROSION RATE, POLARIZATION, AND PASSIVITY

We have seen in Section 10.2 that a current must be passed in order for corrosion to proceed. Conversely, any factor such as cathodic polarization, which limits the current, will limit the cor-

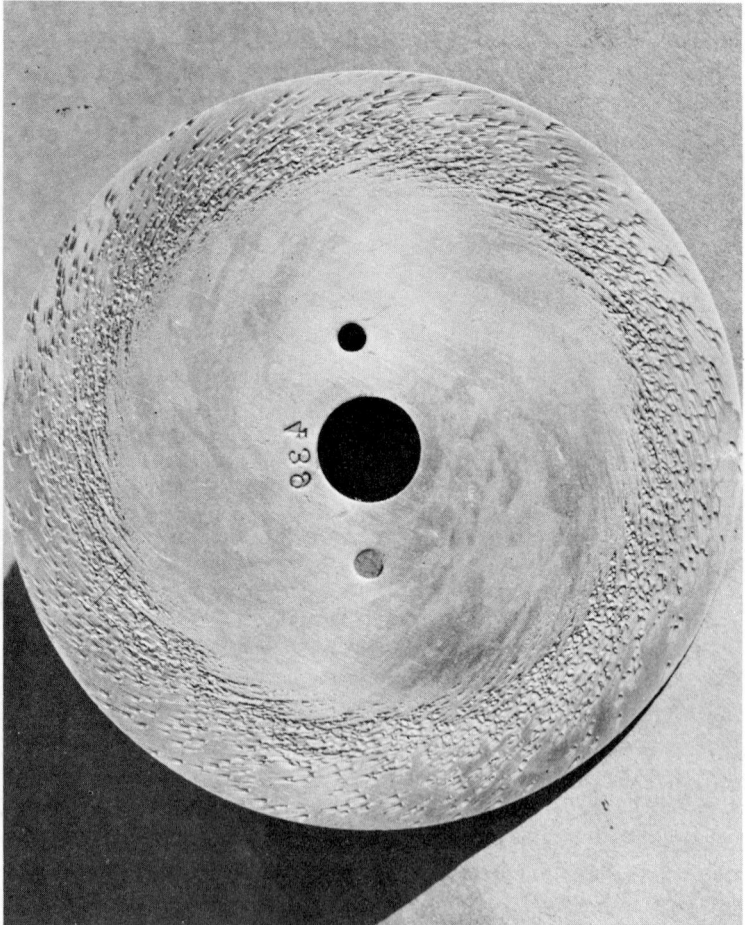

Figure 10.6    This high-speed rotor was attacked where fluid velocity was greatest. Differences in ion concentration due to velocity difference made the outer part anodic. (Courtesy of F. L. La Que, International Nickel Company.)

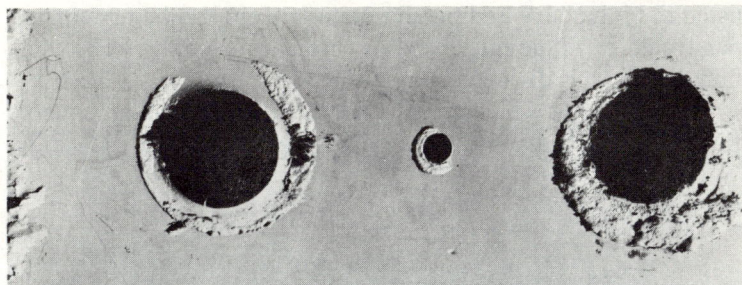

Figure 10.7   Crevice corrosion under a washer on a plate immersed in sea water. The region under the washer was anodic because it lacked oxygen. (Courtesy of F. L. La Que, International Nickel Company.)

rosion rate.   The relative areas of cathode and anode are important since each must carry the same current.   If the cathode area is small, the current *density* at the cathode will be high, relative to that at the anode.   The cathode will therefore polarize rapidly, bringing corrosion to a stop.   Thus a copper nail used to secure steel sheet will lead to little or no corrosion.   If the *anode* is small, however, the small current density at the cathode leads to a low rate of polarization, and attack is rapid.   A steel nail in copper sheet, if used outdoors, rapidly becomes completely converted to rust.   We are fortunate that polarization is so common.   It is easy to see from Section 10.7 that electrochemical potentials are everywhere about us, and only polarization prevents our civilization from collapsing in a heap of corrosion products.

Another important source of protection is the tendency of many metals to form protective, adherent oxide films.   Table 10.1 shows aluminum to be more active than zinc, yet the Galvanic Series (Table 10.2) shows aluminum to be cathodic to zinc in sea water. The corrosion resistance of aluminum is due to a tight, hard, adherent film of oxide on the surface.   Corrosion resistance due to oxide films is called *passivity*.   The oxide film on chromium may be monomolecular but it protects chromium thoroughly in oxidizing environments.   Reducing environments, however, remove the passivating film, and chromium then becomes *active*.   The position in the Galvanic Series depends on whether the metal is active or passive.   In Table 10.2 stainless steel, nickel, and inconel are listed in both active and passive states.  Iron may be temporarily passivated by dilute nitric acid or chromate solutions.  The oxide film, however, is disrupted very easily.

Stainless steels, which are simply alloys of iron containing more than 12% Cr, are passive in any oxidizing atmosphere and exhibit

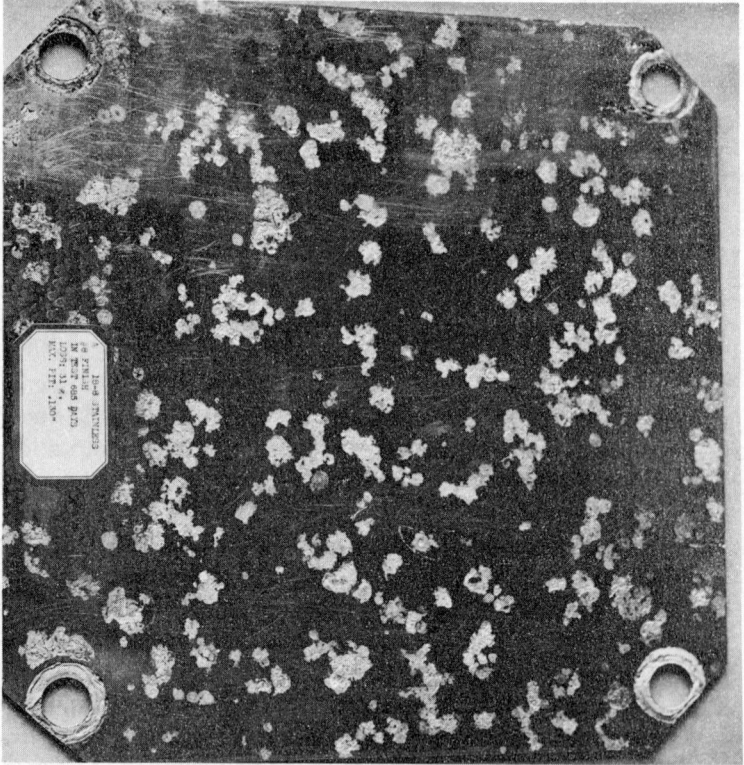

Figure 10.8   Localized corrosion under barnacles on a metal plate in sea water.  The anodic region again was due to lack of oxygen.  (Courtesy of F. L. La Que, International Nickel Company.)

extreme corrosion resistance.  However, in reducing acids such as HCl, stainless steels behave like active iron, and sometimes a good deal worse.  A good example of this behavior is the case of 18:8 stainless (18% Cr, 8% Ni, rest Fe), which has been cooled slowly from 800 to 500°C.  During the cooling chromium carbide pre-cipitates form near the grain boundaries, and the chromium content is depleted below 12% in those regions.  When the chromium content falls below 12%, the material near the grain boundary is no longer "stainless"; it becomes active and anodic to the large

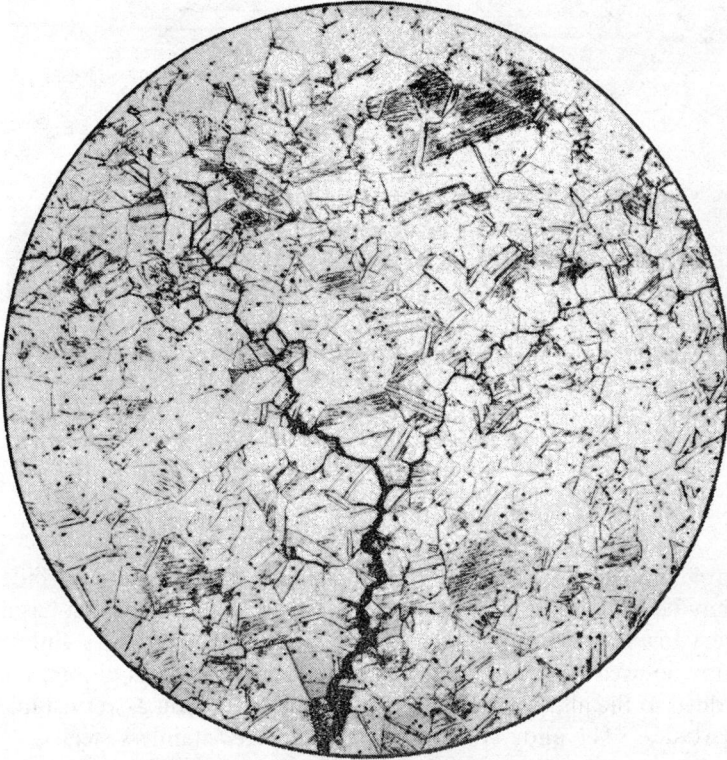

Figure 10.9    Intergranular stress corrosion cracking of brass.  Figure 5, Chapter 19 from *Corrosion and Corrosion Control* by H. H. Uhlig (Wiley, 1963), p. 292.

interior of the grain.  Violent attack near the grain boundaries occurs as shown in Figure 10.10.  Stainless steel which has become susceptible in this way to intergranular corrosion is said to be *sensitized*.  Sensitization may be avoided by rapid cooling.  However, in large structures, and in welded structures (Figure 10.10),

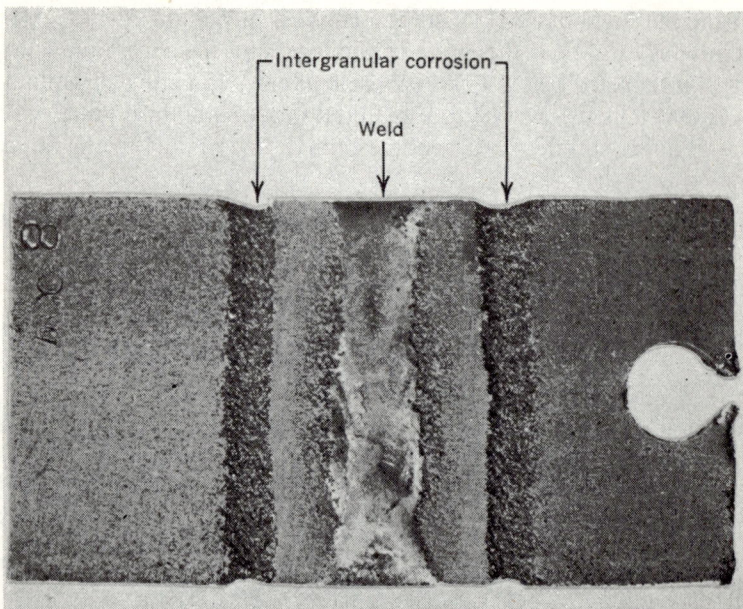

Figure 10.10   Sensitization of stainless steel by welding.   The attacked zone had the proper cooling rate for sensitization.   Figure 2, Chapter 18 from *Corrosion and Corrosion Control* by H. H. Uhlig (Wiley, 1963), p. 267.

rapid cooling is not feasible.   Formation of chromium carbide may be avoided in such cases by using stainless steel which has a very low level of dissolved carbon.   Special low-carbon stainless may be used or, more commonly, an element such as niobium is added to the alloy, which ties up most of the carbon as very stable carbides.   The alloy is then called a *stabilized* stainless steel.

Other passive alloys are 40% Ni–60% Fe and Cu-Ni alloys (such as Monel) containing more than 40% Ni.   All the passive films are broken down by halide ions, and stainless steels and copper-nickel alloys may be pitted severely by sea water.   Conversely, it is possible to protect certain metals by artificial passivation.   Potassium chromate and sodium nitrate both inhibit the corrosion of iron due to passivation.   Red lead and zinc chromate in paint primers for steel and iron act similarly.

## 10.9  CORROSION CONTROL

The most obvious way to minimize corrosion is to avoid all the factors listed in Section 10.7, but this is rarely possible. However, certain practices are feasible. If dissimilar metals are used, electrical insulation may be possible. The cathodic surface area should be kept small. Crevices, recesses, pockets, and sharp corners must be avoided. Good surface finish and relief of internal stresses by heat treatment are desirable. Welded joints, rather than brazed, soldered, or bolted joints, may be used. In addition to all the above practices the following may be used:

(1) Isolation from the environment by inert, organic coatings (paint) or electroplated coatings of noble metals.

(2) Inhibition of cathode or anode reaction by artificial polarization or by agents which react with the corrosion products to leave impervious coatings on electrode surfaces. Such agents are called *inhibitors*. Also, if possible, removal of oxygen from neutral or basic electrolytes to prevent depolarization of the cathode.

(3) The introduction of a more active metal as a *sacrificial anode*. When steel sheet is coated with zinc ("galvanized" sheet) the zinc corrodes. The steel is then the cathode and does not corrode even when the zinc coating is broken or penetrated. Blocks of zinc, bolted to the steel hull of a boat, prevent corrosion of the entire hulls in many cases.

The third technique, called *cathodic protection*, is very effective. The regions of steel which are exposed, as a result of flaws in the zinc coating, polarize rapidly as they are cathodic and of small size. However, if the steel is coated with a more cathodic material such as tin, as in the first technique, the flawed regions are anodic to the very large cathodic regions of tin. Corrosion is rapid. Cathodic protection may also be achieved by connecting the material to the negative terminal of a direct current source. The positive terminal is connected to an expendable electrode. The impressed voltage makes the material of interest cathodic, thereby protecting it. Steel pipelines, buried underground, are protected by using an impressed voltage, with steel scrap as the sacrificial anode.

DEFINITIONS

*Active.* The condition of a metal surface when any protective or passivating films are absent.

*Anode.* The metal in a cell which dissolves as ions and supplies electrons to the external circuit.

*Aqueous Corrosion.* The solution of a metal into an electrolyte in the presence of oxygen and a dissimilar metal.

*Cathode.* The metal in a cell which accepts electrons from the external circuit, and on whose surface hydrogen atoms are neutralized from the solution.

*Cathodic Polarization.* The accumulation of reaction products on the cathode surface to such an extent that the cathode is separated from the electrolyte, and corrosion slows or stops.

*Dissimilar Metals.* Two metals or regions of metal which are chemically unlike because of different composition, state of stress, surface condition, or environment.

*Electromotive Force Series.* An arrangement of metallic elements according to the voltage or electrochemical potential they exhibit in a one-normal solution of ions.

*Galvanic Cell.* Two dissimilar metals in electrical contact in the presence of an aqueous electrolyte.

*Galvanic Protection.* The protection of a metal by connecting it to a sacrificial anode or by impressing a d-c voltage to make it a cathode.

*Galvanic Series.* An arrangement of metallic elements and alloys according to their electrochemical potential in sea water.

*Inhibitors.* Compounds added to an electrolyte which coat the anode or cathode and stop corrosion.

*Intergranular Corrosion.* Preferential attack of grain boundary regions.

*Ion Concentration Cell.* The galvanic cell formed when two pieces of the same metal are electrically connected but are in solutions of different ionic concentration.

*Oxygen Concentration Cell.* The galvanic cell formed when two pieces of the same metal are electrically connected but are in solutions of different oxygen concentration.

*Passivation.* The formation of a film of atoms or molecules on the surface of the anode so that it is effectively separated from the electrolyte and corrosion slows or stops.

*Pit Corrosion.* Rapid, local attack resulting from the formation of small anodic regions at a break in a passive film or corrosion deposit.

*Stress Corrosion.* Preferential attack of areas under stress in a corrosive environment, where this factor alone would not have caused corrosion.

BIBLIOGRAPHY

Birchenall, C. E., *Physical Metallurgy,* McGraw-Hill, New York, 1959.
Jastrzebski, Z. D., *Nature and Properties of Engineering Materials,* Wiley, New

York, 1959.   Both this book and Birchenall's contain excellent undergraduate level chapters on corrosion.

La Que, F. L., *Corrosion in Action,* International Nickel Company, New York, 1955. A simple, attractive booklet containing many excellent illustrations.

Rugger, G. R., "Weathering Resistance of Plastics," *Materials in Design Engineering,* vol. 59, no. 1 (January, 1964), p. 69.

See also:

"Corrosion Resistance Data for Plastics," *Materials in Design Engineering,* vol. 58, no. 1 (January, 1963), p. 106.   Both articles sum up the present state of knowledge on the weathering of plastics in various environments.

Uhlig, H. H., *Corrosion and Corrosion Control,* Wiley, New York, 1963.   A comprehensive textbook on the senior-graduate level.

Uhlig, H. H. (Editor), *Corrosion Handbook,* Wiley, New York, 1948.   A compilation of data and useful knowledge on the subject.

## Problems

10.1   A piece of iron and a piece of copper are placed in a beaker of dilute sodium chloride water solution.   They are externally connected by a wire.   Sketch the cell and indicate:

*a.* Anode location.

*b.* Cathode location.

*c.* Direction of external electron flow.

*d.* Anode half-reaction.

*e.* Cathode half-reaction.

*f.* Which metal corrodes.

10.2   What effect will each of the following have on the cell in Problem 10.1:

*a.* A new dilute HCl electrolyte.

*b.* New $FeSO_4$ electrolyte.

*c.* Bubbling oxygen through the NaCl electrolyte.

*d.* Bubbling nitrogen through the NaCl electrolyte.

*e.* Inserting a battery in the external circuit (two ways).

*f.* Complete de-ionization of the water.

*g.* Disconnecting the wire.

10.3   A cell containing iron connected to copper, immersed in distilled water, exhibits a very slow corrosion rate.   Ferrous sulfate is added and dissolved slowly.   The corrosion rate initially increases and then decreases. Explain why.

10.4   In a tank partially filled with an electrolyte, corrosive attack occurs toward the bottom of the tank.   Explain why.

10.5  Steel pilings immersed in sea water often show corrosive attack near the surface rather than in the depths.  Explain why.

10.6  Explain and contrast the protection afforded to steel by zinc coating (galvanizing) and tin coating.  Remember that all such coatings are likely to contain imperfections, pits, and cracks.

10.7  Explain why stainless steels are particularly susceptible to pitting in sea water.

10.8  Indicate two ways in which buried pipelines can be protected from corrosion.  In what way would local ground conditions influence these methods?

10.9  If the contact of dissimilar metals cannot be avoided, how may the exposed areas of each be varied relative to one another to minimize the rate of the corrosion?

10.10  Assuming all other factors equal, what would be the superior choice to minimize corrosion?

*a.* Cold-worked or annealed metal.

*b.* Single-phase alloy or two phase alloy.

10.11  Explain the difference between the EMF series and the galvanic series of metals.

10.12  Write a short essay on the protection of metals against corrosion by different kinds of paint.

10.13  Why are flash plates of copper followed by nickel recommended as base coats for chromium electroplate?

10.14  (*a*) What is meant by chemisorption?

(*b*) Is the passive film on chromium-containing alloys a chemisorbed film or a true oxide film?

10.15  Describe the kind of defects which might be responsible for initiating pitting in 18:8 stainless steel.

10.16  What metals or alloys lend themselves best for bone inserts and sutures in surgery?  Discuss the relative merits and shortcomings of the ones now used.

10.17  How are magnesium castings usually protected against atmospheric corrosion?

# *Tabular Data*

The tables which follow contain a collection of useful data in these fields:

*Table A.1    Summary of Thermodynamic Relations[a]*

| $X$ | $Y$ | $Z$ | $\left(\dfrac{\partial Y}{\partial X}\right)_Z$ | $X$ | $Y$ | $Z$ | $\left(\dfrac{\partial Y}{\partial X}\right)_Z$ |
|---|---|---|---|---|---|---|---|
| $T$ | $V$ | $P$ | $\alpha V$ | $T$ | $P$ | $V$ | $\alpha/\beta$ |
| $T$ | $S$ | $P$ | $C_P/T$ | $T$ | $S$ | $V$ | $C_P/T - \alpha^2 V/\beta$ |
| $T$ | $V$ | $P$ | $C_P - \alpha PV$ | $T$ | $E$ | $V$ | $C_P - \alpha^2 VT/\beta$ |
| $T$ | $H$ | $P$ | $C_P$ | $T$ | $H$ | $V$ | $C_P - \alpha^2 VT/\beta + \alpha V/\beta$ |
| $T$ | $F$ | $P$ | $-\alpha PV - S$ | $T$ | $F$ | $V$ | $-S$ |
| $T$ | $G$ | $P$ | $-S$ | $T$ | $G$ | $V$ | $\alpha V/\beta - S$ |
| $P$ | $V$ | $T$ | $-\beta V$ | $T$ | $P$ | $S$ | $C_P/\alpha VT$ |
| $P$ | $S$ | $T$ | $-\alpha V$ | $T$ | $V$ | $S$ | $-\beta C_P/\alpha T + \alpha V$ |
| $P$ | $E$ | $T$ | $\beta PV - \alpha VT$ | $T$ | $E$ | $S$ | $\beta PC_P/\alpha T - \alpha PV$ |
| $P$ | $H$ | $T$ | $V - \alpha VT$ | $T$ | $H$ | $S$ | $C_P/\alpha T$ |
| $P$ | $F$ | $T$ | $\beta PV$ | $T$ | $F$ | $S$ | $\beta PC_P/\alpha T - \alpha PV - S$ |
| $P$ | $G$ | $T$ | $V$ | $T$ | $G$ | $S$ | $C_P/\alpha T - S$ |

[a] From J. Lumsden, *Thermodynamics of Alloys,* Institute of Metals, London, 1952.


*Table A.2    Heat Capacity Constants for Number of Substances[a]*

| Substance | $a$ | $b \times 10^3$ | $c \times 10^{-5}$ | Temperature Range (°K) |
|---|---|---|---|---|
| $Ag(s)$ | 5.09 | 2.04 | 0.36 | 298–m.p. |
| $Ag(l)$ | 7.30 | — | — | m.p.–1600 |
| $AgBr(s)$ | 7.93 | 15.40 | — | 298–m.p. |
| $AgCl(s)$ | 14.88 | 1.00 | $-2.70$ | 298–m.p. |
| $BCl(l)$ | 16.86 | 2.86 | $-2.44$ | 298–1000 |
| $CaTiSiO_5(s)$ | 42.39 | 5.54 | $-9.63$ | 298–1670 |
| $CaTiSiO_5(l)$ | 66.8 | — | — | 1670–1811 |
| $SiO_2(\text{quartz})$ | 11.22 | 8.20 | $-2.70$ | 298–848 |
| $MnSiO_3(s)$ | 26.42 | 3.88 | $-6.16$ | 298–1500 |

[a] From O. Kubaschewski and E. Evans, *Metallurgical Thermochemistry,* Third Edition, Pergamon, London, 1958.

*Table A.3   Absolute Entropies of Some Substances*[a]

| Substance | $S^{\circ}_{298}$(cal/mole degree) | Substance | $S^{\circ}_{298}$(cal/mole degree) |
|---|---|---|---|
| Ag($s$) | 10.20 ± 0.05 | Graphite($s$) | 1.361 ± 0.005 |
| AgCl($s$) | 23.0 ± 0.1 | H($g$) | 27.4 ± 0.01 |
| Al($s$) | 6.77 ± 0.05 | H$_2$($g$) | 31.21 ± 0.02 |
| Au($s$) | 11.32 ± 0.05 | MgO($s$) | 6.55 ± 0.15 |
| B($s$) | 1.4 ± 0.05 | MnO($s$) | 14.3 ± 0.2 |
| BCl$_3$($l$) | 49.2 ± 2.0 | MnSiO$_3$($s$) | 21.3 ± 0.2 |
| BCl$_3$($s$) | 69.3 ± 0.5 | O$_2$($g$) | 49.02 ± 0.01 |
| BeO | 3.37 ± 0.05 | Si($s$) | 4.5 ± 0.05 |
| Diamond($s$) | 0.583 ± 0.005 | Sn(gray) | 10.7 ± 0.1 |
| Fe($s$) | 6.49 ± 0.03 | Sn(white) | 12.3 ± 0.1 |
| Ge($s$) | 10.1 ± 0.2 | | |

[a] From O. Kubaschewski and E. Evans, *Metallurgical Thermochemistry*, Third Edition, Pergamon, London, 1958.

Table A.4  *Standard Free Energies of Reactions*[a]

$$\Delta G^\circ = a + bT \log T + cT \text{ (in cal)}$$

| Reaction | $a$ | $b$ | $c$ | $\pm$Kcal | Temperature Range (°K) |
|---|---|---|---|---|---|
| $2Al(s) + \frac{3}{2}O_2(g, 1\ atm) = Al_2O_3(s)$ | $-400{,}810$ | $-3.98$ | $87.64$ | 3 | 298–923 |
| $2Al(l) + \frac{3}{2}O_2(g, 1\ atm) = Al_2O_3(s)$ | $-405{,}760$ | $-3.75$ | $92.22$ | 4 | 923–1800 |
| $2Be(s) + O_2(g) = 2BeO$ | $-286{,}900$ | $-3.32$ | $56.1$ | 10 | 298–1557 |
| $Ca(s) + Si(s) = CaSi(s)$ | $-36{,}000$ | — | $-0.5$ | 4 | 298–1123 |
| $2CaO(s) + SiO_2(s) = Ca_2SiO_4$ | $-30{,}200$ | — | $-1.2$ | 2.5 | 298–1700 |
| $2Co(s) + C = Co_2C(s)$ | $3{,}950$ | — | $-2.08$ | 5 | 298–1200 |
| $Fe(s) + \frac{1}{2}O_2(g) = FeO(s)$ | $-62{,}050$ | — | $14.95$ | 3 | 298–1642 |
| $\frac{1}{4}Ge(s) + \frac{1}{2}GeO(g) = GeO(g, 1\ atm)$ | $-54{,}600$ | $-6.9$ | $62.0$ | 4 | 298–860 |
| Graphite = diamond | $-310$ | — | $1.13$ | 0.2 | 298–1500 |
| $La(s) + \frac{1}{2}N_2(g) = LaN(s)$ | $-72{,}100$ | — | $25.0$ | 9 | 298–1000 |
| $2MgO(s) + SiO_2(s) = Mg_2SiO_4(s)$ | $-15{,}120$ | — | $0.0$ | 2 | 298–1700 |
| $MnO(s) + SiO_2(s) = MnSiO_3(s)$ | $-5{,}920$ | — | $3.0$ | 4 | 298–1600 |
| $Mo(s) + O_2(g) = MoO_2(s)$ | $-140{,}100$ | $-4.6$ | $55.8$ | 6 | 298–1300 |
| $Na_2O(s) + SiO_2(s) = Na_2SiO_3$ | $-55{,}550$ | — | $1.40$ | 8 | 298–1361 |
| $Ti(s) + 2Cl_2(g, 1\ atm) = TiCl_4(g, 1\ atm)$ | $-180{,}700$ | $-1.8$ | $34.65$ | 3 | 298–1700 |
| $Si(s) + O_2(g, 1\ atm) = SiO_2(s)$ | $-210{,}600$ | $-3.0$ | $52.22$ | 3 | 298–1700 |

[a] From Kubaschewski and Evans, *Metallurgical Thermochemistry*, Third Edition, Pergamon, London, 1958.

182

*Table A.5*   *Relation between Melting Point and Thermal Expansion*[b]

| Structure | Metal | M.P. | $\alpha \times 10^{-6}$ cm/cm/°C | $\alpha \times$ M.P. $\times 10^3$ |
|---|---|---|---|---|
| FCC | Cu | 1356 | 16.5 | 22.5 |
| | Ag | 1233 | 19.7 | 18.9 |
| | Au | 1336 | 16.2 | 18.8 |
| | Pt | 2046 | 8.9 | 18.3 |
| | Ir | 2727 | 6.8 | 18.6 |
| | Rh | 2239 | 8.3 | 18.3 |
| | Pd | 1827 | 11.8 | 21.5 |
| | Al | 933 | 23.9 | 22.3 |
| | Ni | 1728 | 13.3 | 23.0 |
| | Ca[a] | 1123 | 22.0 | 25.0 |
| | Pb | 600 | 29.3 | 17.6 |
| | Th | 2073 | 11.1 | 21.2 |
| CP Hex | Cd | 594 | 29.8 | 17.8 |
| | Zn | 692 | 15.0 | 10.4 |
| | | | 61.8 | 42.5 |
| | Mg | 923 | 26.0 | 23.1 |
| | Be | 1553 | 12.4 | 15.8 |
| | Co | 1768 | 12.3 | 21.7 |
| | Os | 2973 | 4.6 | 13.7 |
| | Te[a] | 573 | 11.1 | 21.2 |
| | Ti[a] | 2193 | 8.5 | 18.6 |
| | Zr[a] | 2023 | 5.0 | 10.2 |
| BCC | Li | 459 | 56 | 25.6 |
| | Na | 371 | 71 | 26.2 |
| | K | 336 | 83 | 27.7 |
| | Rb | 312 | 90 | 28.0 |
| | Cs | 301 | 97 | 29.2 |
| | V | 2005 | 7.8 | 15.7 |
| | Cr | 2163 | 6.2 | 13.2 |
| | Fe[a] | 1612 | 11.7 | 18.8 |
| | Cb | 2688 | 71 | 19.1 |
| | Mo | 2898 | 4.9 | 14.2 |
| | Ta | 3269 | 6.5 | 21.2 |
| | W[a] | 3783 | 4.3 | 16.3 |
| Other | Sb | 903 | 9.0 | 8.1 |
| | Bi | 793 | 13.3 | 10.5 |
| | Ga | 303 | 18.0 | 5.5 |
| | In | 429 | 33.0 | 14.1 |
| | Sn | 504 | 23.0 | 11.6 |
| | Mn[a] | 1518 | 22.0 | 33.5 |

[a] Denotes allotropic change between room temperature and the melting point.
[b] From B. Chalmers, *Physical Metallurgy*, Wiley, New York, 1959.

*Table A.6    The Effect of Pressure on the Melting Point of Solids[a]*

| Substance | $T_{m.p.}$ ($°K$) | $\Delta H_f$ (cal/gm) | $\Delta V$ ($V_l - V_s$) ($cm^3$/gm) | $\Delta T$ for 1000 atm Calc | $\Delta T$ for 1000 atm Obs |
|---|---|---|---|---|---|
| $H_2O$ | 273.2 | 79.8 | −0.0906 | −7.5 | −7.4 |
| Acetic acid | 289.8 | 44.7 | +0.01595 | +25.0 | +24.4 |
| Tin | 505.0 | 14.0 | +0.00389 | +3.40 | +3.28 |
| Bismuth | 544.0 | 12.6 | −0.00342 | −3.56 | −3.55 |

[a] From H. M. Strong, *Am. Scientist*, **48**, 58 (1960).

*Table A.7    Vapor Pressure Constants of Several Substances (in mm Hg)[a]*

| Substance | $a$ | $b$ | $c \times 10^3$ | $d$ | Temperature Range ($°K$) |
|---|---|---|---|---|---|
| Ag($s$) | −14,710 | −0.755 | — | 11.66 | 298–m.p. |
| Ag($l$) | −14,260 | −1.055 | — | 12.23 | m.p.–b.p. |
| BeO($s$) | −34,230 | −2.0 | — | 18.50 | 298–m.p. |
| Ge($s$) | −20,150 | −0.91 | — | 13.28 | 298–m.p. |
| Mg($s$) | −7,780 | −0.855 | — | 11.41 | 298–m.p. |
| Mg($l$) | −7,550 | −1.41 | — | 12.79 | m.p.–b.p. |
| NaCl($s$) | −12,440 | −0.90 | −0.46 | 14.31 | 298–m.p. |
| Si($s$) | −18,000 | −1.022 | — | 12.83 | 1200–m.p. |

[a] From Kubaschewski and Evans, *Metallurgical Thermochemistry*, Third Edition, Pergamon, London, 1958.

Table A.8    Latent Heat of Fusion as a Function of the Absolute
Melting Temperatures of Various Elements[a]

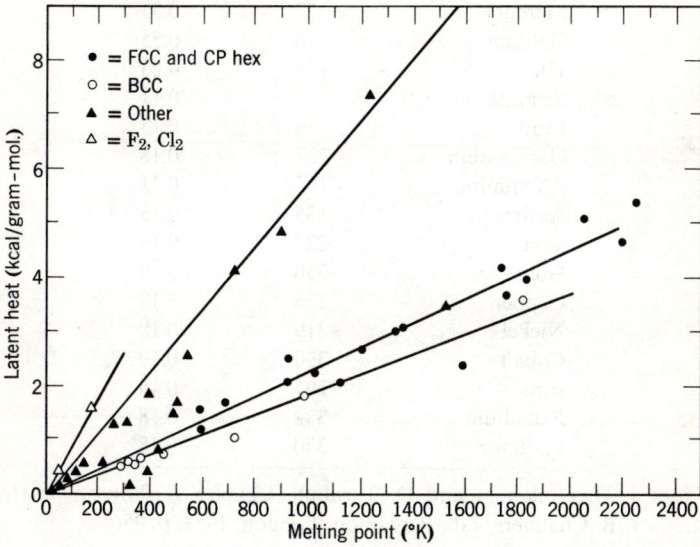

[a] From B. Chalmers, *Physical Metallurgy,* Wiley, New York, 1959.

### Table A.9  Supercooling for Homogeneous Nucleation[a]

| Metal | Supercooling ($\Delta H$) | $\dfrac{\Delta H}{T_E}$ |
|---|---|---|
| Mercury | 77 | 0.33 |
| Gallium | 76 | 0.25 |
| Tin | 118 | 0.23 |
| Bismuth | 90 | 0.17 |
| Lead | 80 | 0.13 |
| Germanium | 227 | 0.18 |
| Aluminum | 195 | 0.21 |
| Antimony | 135 | 0.15 |
| Silver | 227 | 0.18 |
| Gold | 230 | 0.17 |
| Copper | 236 | 0.17 |
| Nickel | 319 | 0.18 |
| Cobalt | 330 | 0.19 |
| Iron | 295 | 0.16 |
| Palladium | 332 | 0.18 |
| Platinum | 370 | 0.18 |

[a] From J. H. Hollomon and D. Turnbull, Chapter 7, *Progress in Metal Physics 4*, B. Chalmers, Ed., Pergamon, London, 1953, p. 356.

### Table A.10  Change of Volume on Melting[a]

| Element | Increase in Volume | Element | Increase in Volume |
|---|---|---|---|
| Li | 1.65 | Hg | 3.7 |
| Na | 2.5 | Al | 6.0 |
| K | 2.55 | Ga | −3.2 |
| Rb | 2.5 | Te | 3.2 |
| Cs | 2.6 | Si | −12.0 |
| | | Ge | −12.0 |
| Cu | 4.15 | Sn | 2.8 |
| Ag | 3.8 | Pb | 3.5 |
| Au | 5.1 | Sb | −0.95 |
| | | Bi | −3.35 |
| Mg | 4.1 | | |
| Zn | 4.2 | | |
| Cd | 4.7 | | |

[a] From B. R. T. Frost, Chapter 3, *Progress in Metal Physics 5*, B. Chalmers and R. King, Eds., Pergamon, London, 1954, p. 98.

## Table A.11 Values of Free Energy, Enthalpy, and Entropy of Solid-Gas Interfaces

| Metal | $\Delta G^{SV}$ (ergs/cm$^2$) | $\Delta H^{SV}$ (ergs/cm$^2$) | $\Delta S^{SV}$ (ergs/cm$^2$ °K) | $T(°C)$ |
|---|---|---|---|---|
| Ag[a] | 1140 | 1678 | 0.47 | 903 |
| Au[b] | 1400 | 2006 | 0.43 | 1204 |
| Cu[c] | 1650 | 2350 | 0.55 | 1000 |

[a] From E. R. Funk, H. Udin, and J. Wulff, *J. Metals*, **3**, 1206 (1951).
[b] From F. H. Buttner, H. Udin, and J. Wulff, *J. Metals*, **3**, 1209 (1951).
[c] From H. Udin, A. J. Shaler, and J. Wulff, *J. Metals*, **1**, 1936 (1949).

## Table A.12 Relationship between Solid-Vapor Surface Energy and Latent Heat of Vaporization[a]

| Metal | Solid-Vapor Surface Energy (ergs/cm$^2$) | Latent Heat Vaporization (kcal/mole) | Surface Energy/Atom Latent Heat/Atom |
|---|---|---|---|
| Copper | 1700 | 73.3 | 0.14 |
| Silver | 1200 | 82.0 | 0.26 |
| Gold | 1400 | 60.0 | 0.18 |

[a] From B. Chalmers, *Physical Metallurgy*, Wiley, New York, 1959.

## Table A.13 Relative Interface Free Energies[a]

| System | Interface between Phase A | Phase B | Grain Boundary Used as a Comparison Interface, $D$ | $\dfrac{\Delta G_{AB}{}^{b}}{\Delta G_{D}{}^{b}}$ | $T(°C)$ |
|---|---|---|---|---|---|
| Cu-Zn | α FCC | β BCC | α/α | 0.78 | 700 |
| Cu-Zn | α FCC | β BCC | β/β | 1.00 | 700 |
| Cu-Al | α FCC | β BCC | α/α | 0.71 | 600 |
| Cu-Al | β BCC | γ C.C.[c] | γ/γ | 0.78 | 600 |
| Cu-Sn | α FCC | β BCC | α/α | 0.76 | 750 |
| Cu-Sb | α FCC | β BCT | α/α | 0.71 | 600 |
| Cu-Ag | α (Cu) FCC | β (Ag) FCC | β/β | 0.74 | 750 |
| Cu-Si | α FCC | β BCC | α/α | 0.53 | 845 |
| Cu-Si | α FCC | β BCC | β/β | 1.18 | 845 |
| Fe-C | α BCC | Fe$_3$C o.r.[c] | α/α | 0.93 | 690 |
| Fe-C | α BCC | γ FCC | α/α | 0.71 | 750 |
| Fe-C | α BCC | γ FCC | α/α | 0.74 | 950 |
| Fe-Cu | α BCC | FCC | α/α | 0.74 | 825 |
| Zn-Sn | β (Sn) BCT | α (Zn) HCP | α/α | 0.74 | 160 |

[a] From C. S. Smith, *Imperfections in Nearly Perfect Crystals*, W. Shockley, Ed., Wiley, New York, 1952.
[b] From C. S. Smith, *Trans. A.I.M.E.*, **175**, 15, 1948.
[c] C.C. = Complex Cubic; o.r. = orthorhombic.

*Table A.14    Values of Surface Free Energies for Compounds*

| Compound | $\Delta G^{SV}$ (ergs/cm$^2$) | Compound | $\Delta G^{SV}$ (ergs/cm$^2$) |
|---|---|---|---|
| NaCl[a] (100) | 300 | CaF$_2$[a] (111) | 450 |
| LiF[b] (100) | 340 | BaF$_2$[b] (111) | 280 |
| MgO[b] (100) | 1200 | CaCO$_3$[b] (1010) | 230 |

[a] From U. D. Kusnetov and P. P. Feterim, *Surface Energy of Solids,* Her Majesty's Stationery Office, London, 1957.
[b] From J. J. Gilman, *J. Appl. Phys.,* **31,** 2208 (1960).

*Table A.15    Comparison of Calculated and Experimental Values of Surface Energies[a]*

| Crystal | Experimental $\Delta G^{SV}$ (ergs/cm$^2$) | Calculated $\Delta G^{SV}$ (ergs/cm$^2$) |
|---|---|---|
| NaCl | 300 | 310 |
| LiF | 340 | 370 |
| MgO | 1200 | 1300 |
| CaF$_2$ | 450 | 540 |
| BaF$_2$ | 280 | 350 |
| CaCO$_3$ | 230 | 380 |
| Si | 1240 | 890 |
| Zn | 105 | 185 |
| Fe (3% Si) | 1360 (?) | 1400 |

[a] From J. J. Gilman, *J. Appl. Phys.,* **31,** 2208 (1960).

*Table A.16    Interfacial Energies for High Angle Grain Boundaries and Relation to Surface Energy[a]*

| Crystal | $\Delta G^b$ (ergs/cm$^2$) | $\Delta G^b/\Delta G^{SV}$ | Reference |
|---|---|---|---|
| Copper | 600 | 0.36 | N. A. Gjostein and F. N. Rhines, *Acta Met.,* **7,** 319 (1959). |
| $\gamma$-Iron | 850 | — | L. H. Van Vlack, *J. Metals,* **3,** 25 (1951). |
| $\alpha$-Iron (4% Si) | 760 | 0.55 | L. H. Van Vlack, *ibid.* |
| Lead | 200 | — | K. T. Aust and B. Chalmers, *Proc. Roy. Soc. (London),* **A204,** 359 (1951). |
| Tin | 100 | — | K. T. Aust and B. Chalmers, *Proc. Roy. Soc. (London),* **A201,** 210 (1950). |
| Silver | 400 | 0.35 | A. P. Greenough and R. King, *J. Inst. Met.,* **79,** 415 (1951). |

[a] From R. A. Swalin, *Thermodynamics of Solids,* Wiley, New York, 1962.

*Table A.17   Comparison of Interphase and Intraphase Boundary Energies*[a]

| System | Interface between Phase A | Phase B | Comparison Grain Boundary ($c$) | Energies $\dfrac{\gamma AB}{\gamma c}$ |
|---|---|---|---|---|
| CuZn | α (FCC) | β (BCC) | α/α | 0.78 |
| CuZn | α (FCC) | β (BCC) | β/β | 1.00 |
| CuAl | α (FCC) | β (BCC) | α/α | 0.71 |
| CuAl | β (BCC) | γ (Complex cubic γ brass) | γ/γ | 0.78 |
| CuSn | α (FCC) | β (BCC) | α/α | 0.76 |
| CuSn | α (FCC) | β (BCC) | β/β | 0.93 |
| FeC | α (BCC) | $Fe_3C$ (Orthorhombic) | α/α | 0.93 |
| FeC | α (BCC) | γ (FCC) | α/α | 0.71 |
| FeC | α (BCC) | γ (FCC) | γ/γ | 0.74 |

[a] From C. S. Smith, *Imperfections in Nearly Perfect Crystals,* Shockley, Ed., Wiley, New York, 1952, p. 384.

*Table A.18   Solid-Liquid Interfacial Energy and Comparison with Atomic Latent Heat*[a]

| Metal | Solid-Liquid Interfacial Energy ($ergs/cm^2$) | Interfacial Energy/Atom / Latent Heat/Atom |
|---|---|---|
| Mercury | 24.4 | 0.53 |
| Gallium | 56 | 0.44 |
| Tin | 54.5 | 0.42 |
| Bismuth | 54 | 0.33 |
| Lead | 33 | 0.39 |
| Antimony | 101 | 0.30 |
| Germanium | 181 | 0.35 |
| Silver | 126 | 0.46 |
| Gold | 132 | 0.44 |
| Copper | 177 | 0.44 |
| Manganese | 206 | 0.48 |
| Nickel | 255 | 0.44 |
| Cobalt | 234 | 0.40 |
| Iron | 204 | 0.45 |
| Palladium | 209 | 0.45 |
| Platinum | 240 | 0.46 |

[a] From J. H. Hollomon and D. Turnbull, Chapter 7, *Progress in Metal Physics 4,* B. Chalmers, Ed., Pergamon, London, 1953.

*Table A.19    Relative Energies of Solid-Liquid Interfaces
and Grain Boundary Energies[a]*

| System | Solid Grain Boundary | $\dfrac{\gamma SL}{\gamma SS}$ |
|---|---|---|
| CuPb | $\alpha/\alpha$ (FCC) | 0.58 |
| CuZnPb | $\alpha/\alpha$ (30% Zn, FCC) | 0.65 |
| CuZnPb | $\beta/\beta$ (49% Zn, BCC) | 0.87 |
| FeCu | $\gamma/\gamma$ (FCC) | 0.51 |
| AlSn | Al/Al (FCC) | 0.56 |
| FeAg | $\gamma/\gamma$ (FCC) | >4 |
| FeFeS | $\gamma/\gamma$ (FCC) | 0.52 |

[a] From C. S. Smith, *Imperfections in Nearly Perfect Crystals,* Shockley, Ed.,
Wiley, New York, 1952, p. 385.

*Table A.20    Diffusion of Carbon in Iron[a]*

| | Concentration in Weight per Cent | | Temperature °C | $D$ (cm²/sec) | |
|---|---|---|---|---|---|
| | Electrolyte-Fe | | 925 | $1.2 \times 10^{-7}$ | |
| Gaseous carburizing agent | C = 0.07 P = 0.003 | | 925 | $3.0 \times 10^{-7}$ | |
| | Mn = 0.27 Si, Si traces | | 1000 | $1.93 \times 10^{-6}$ | |
| Gaseous carburizing agent | Armco-Fe C = 0.02 | | 800 | $1.5 \times 10^{-8}$ | |
| | | | 900 | $7.5 \times 10^{-8}$ | |
| | Si = 0.02 | | 950 | $1.18 \times 10^{-7}$ | |
| | Mn = 0.05 | | 1000 | $2.0 \times 10^{-7}$ | |
| | S = 0.03 | | 1050 | $2.8 \times 10^{-7}$ | |
| | P = 0.012 | | 1100 | $4.5 \times 10^{-7}$ | |
| | | | | 1.0% C | 1.5% C |
| Decarburization of white cast iron in $CO$–$CO_2$-mixture | C = 1.82—3.16 | | 950 | $1.17 \times 10^{-7}$ | $1.3 \times 10^{-7}$ |
| | Si = 0.33—0.45 | | 1000 | $2.83 \times 10^{-7}$ | $2.88 \times 10^{-7}$ |
| | Mn = 0.36—0.52 | | 1050 | $4.54 \times 10^{-7}$ | $5.27 \times 10^{-7}$ |
| | P = 0.067—0.143 | | 1100 | $8.3 \times 10^{-7}$ | $7.1 \times 10^{-7}$ |
| | High carbon steel | Armco-Fe | | | |
| Welding of high carbon steel with Armco iron | C = 1.10 | 0.030 | 925 | $1.08 \times 10^{-7}$ | |
| | Si = 0.282 | 0.005 | 1000 | $2.7 \times 10^{-7}$ | |
| | Mn = 0.230 | 0.027 | 1100 | $7.23 \times 10^{-7}$ | |
| | P = 0.014 | 0.012 | 1200 | $1.95$—$2.25 \times 10^{-6}$ | |
| | S = 0.006 | 0.028 | 1250 | $2.8 \times 10^{-6}$ | |

[a] From W. Jost, *Diffusion in Solids, Liquids, Gases,* Academic, New York, 1952.

Table A.21    Diffusion of Metals in Iron[a]

| Diffusing Metal | Initial Concentration in Atom per Cent | Temperature °C | $D$ (cm²/sec) |
|---|---|---|---|
| Al | Pure | 900 | 3.8 $\times 10^{-9}$ |
|    |      | 1050 | 2.0 $\times 10^{-8}$ |
| Cr | Pure | 1150 | 6.8 $\times 10^{-10}$ |
|    |      | 1300 | 2.2–5.3 $\times 10^{-8}$ |
|    | Fe-Cr-powder | 1200 | 1.7–8.1 $\times 10^{-9}$ |
| Mn | ~27 | 960 | 3.0 $\times 10^{-10}$ |
|    | 3 | 1400 | 9.6 $\times 10^{-8}$ |
| Mo | 0–3.1 | 1200 | 2.3–3.0 $\times 10^{-9}$ |
| Ni | 22 | 1200 | 9.3 $\times 10^{-11}$ |
| Si | 35 | 960 | 7.5 $\times 10^{-9}$ |
|    |    | 1150 | 1.45 $\times 10^{-8}$ |
| Sn | Pure | 950 | 9.7 $\times 10^{-10}$ |
|    |      | 1000 | 2.0 $\times 10^{-9}$ |
|    |      | 1050 | 3.9 $\times 10^{-9}$ |
|    |      | 1100 | 7.6 $\times 10^{-9}$ |
| W | 0–1.3 | 1280 | 3.7 $\times 10^{-10}$ |
|   | 0–1.2 | 1330 | 2.4 $\times 10^{-9}$ |
|   | 0–3.4 | 1330 | 1.0 $\times 10^{-8}$ |

[a] Adapted from W. Jost, Diffusion in Solids, Liquids, Gases, Academic, New York, 1952.

Table A.22　Relation Between Activation Energy of Self-Diffusion Q, Melting Point $T_m$, and Heat of Sublimation L, $D = D_0 \exp(-Q/RT)$

| Metal | $D_0$ (cm²/sec) | $Q$ (kcal/g atom) | $T_m$ °K | $Q/T_m$ | $L$ (kcal) | $Q/L$ |
|---|---|---|---|---|---|---|
| Ag | 0.895 | 45.9 | 1234 | 37.0 | 68.0 | 0.68 |
| Au | 0.16 | 53.0 | 1336 | 40.0 | 92.0 | 0.58 |
|  | 0.02 | 51.0 | 1336 | 38.0 | 92.0 | 0.55 |
| Cu | 47.0 | 61.4 | 1356 | 45.0 | 81.2 | 0.76 |
|  | 11.0 | 57.2 | 1356 | 42.0 | 81.2 | 0.70 |
| α-Fe | $3.4 \times 10^4$ | 77.2 | 1803 | 43.0 | 96.0 | 0.80 |
| γ-Fe | $1.04 \times 10^{-3}$ | 48.0 | 1803 | 26.7 |  |  |
| Pb | 6.7 | 27.9 | 600 | 46.5 | 47.5 | 0.60 |
| β-W |  | 140.0 | 3673 | 38.0 | 203.0 | 0.69 |
|  |  |  | Nonregular metals |  |  |  |
| Bi ∥ c | $1.2 \times 10^{-3}$ | 31.0 | 554 | 57.0 | 47.8 | 0.65 |
| ⊥ c | $6.9 \times 10^{46}$ | 140.0 | 554 | 257.0 | 47.8 | 2.92 |
| Zn ∥ c | $4.6 \times 10^{-2}$ | 20.4 | 693 | 29.5 | 27.4 |  |
| ⊥ c | 92.0 | 31.0 | 693 | 44.7 | 27.4 |  |

[a] Not self-diffusion but diffusion of Fe extrapolated to zero concentration.
From W. Jost, Diffusion in Solids, Liquids, Gases, Academic, New York, 1952.

Table A.23　Diffusion Constants in Metals $D = D_0 \exp(-Q/RT)$

| Solvent Metal | Diffusing Metal | Concentration[a] (Atom per Cent) | Temperature Interval °C | $D_0$ (cm²/sec) | $Q$ (kcal/g atom) |
|---|---|---|---|---|---|
| Ag | Ag | (Self-diff.) | 725–950 | 0.895 | 45.95 |
| Ag | Au | Pure Au | 218–601 | $5.3 \times 10^{-4}$ | 29.8 |
| Ag | Au | 18.4 | 767–916 | $1.1 \times 10^{-4}$ | 26.6 |
| Ag | Cd | 2.0 | 650–895 | $4.9 \times 10^{-5}$ | 22.35 |
| Ag | Cu | 2.0 | 650–895 | $5.9 \times 10^{-5}$ | 24.8 |
| Ag | In | 2.0 | 650–895 | $7.3 \times 10^{-5}$ | 24.4 |
| Ag | Pd | 20.2 | 444–917 | $6.4 \times 10^{-6}$ | 20.2 |
| Ag | Sb | 2.0 | 650–895 | $5.3 \times 10^{-5}$ | 21.7 |
| Ag | Sn | 2.0 | 650–895 | $7.8 \times 10^{-5}$ | 21.4 |

[a] Initial concentration of alloy.

*Table A.23* (continued)

| Solvent Metal | Diffusing Metal | Concentration[a] (Atom per Cent) | Temperature Interval °C | $D_0$ (cm²/sec) | $Q$ (kcal/g atom) |
|---|---|---|---|---|---|
| Al | Ag | 1.26 | 466–573 | 1.1 | 32.6 |
| Al | Cu | Eutectic | 440–540 | 2.3 | 34.9 |
| Al | Cu | 0.85 or 0.17 | 457–565 | $8.4 \times 10^{-2}$ | 32.6 |
| Al | Mg | Eutectic | 365–440 | $1.5 \times 10^{-2}$ | 38.5 |
| Al | Mg | 5.5–11.0 | 395–577 | $1.2 \times 10^{-1}$ | 28.6 |
| Al | Mg | 5.3–8.0 | 420–520 | | 38.0 |
| Al | Si | 0.50 | 465–600 | $9.0 \times 10^{-1}$ | 30.55 |
| Al | Zn | 0.84 | 415–555 | $1.2 \times 10^{1}$ | 27.8 |
| Au | Au | (Self-diff.) | 800–1020 | 0.16 | 53.0 |
| Au | Au | (Self-diff.) | 721–966 | $2 \times 10^{-2}$ | 51.0 |
| Au | Ag | 9.0 | 850–1000 | $2.9 \times 10^{-2}$ | 38.0 |
| Au | Cu | Pure Cu | 301–616 | $1.06 \times 10^{-3}$ | 27.4 |
| Au | Cu | 25.6 | 443–740 | $5.8 \times 10^{-4}$ | 27.4 |
| Au | Fe | 18.3 | 753–1003 | $1.16 \times 10^{-4}$ | 24.4 |
| Au | Ni | 15.0 | 800–1003 | $1.74 \times 10^{-3}$ | 31.2 |
| Au | Pd | 17.1 | 727–970 | $1.13 \times 10^{-3}$ | 37.4 |
| Au | Pt | 20.1 | 740–986 | $1.24 \times 10^{-3}$ | 39.0 |
| Bi | Bi | ‖ c (Self-diff.) | 212–269 | $\sim 10^{-3}$ | 31.0 |
| Bi | Bi | ⊥c(Self-diff.) | 209–269 | $\sim 10^{47}$ | 140.0 |
| Cd | Hg | 4.0 | 156–202 | 2.6 | 19.6 |
| Cu | Cu | (Self-diff.) | 830–1030 | 47.0 | 61.4 |
| Cu | Cu | (Self-diff.) | 750–950 | 11.0 | 57.2 |
| Cu | Ag | 3.0 | 720–860 | $2.9 \times 10^{-2}$ | 37.2 |
| Cu | Al | 15–21 | 500–850 | $7.1 \times 10^{-2}$ | 39.2 |
| Cu | Au | 2.4–3.5 | 400–970 | $6.8 \times 10^{-6}$ | 22.5 |
| Cu | Cd | 3.0 | 720–860 | $3.04 \times 10^{-4}$ | 23.7 |
| Cu | Mn | 8.0–11.4 | 400–850 | $7.2 \times 10^{-6}$ | 23.2 |
| Cu | Ni | 7.5–11.8 | 550–950 | $6.5 \times 10^{-5}$ | 29.8 |
| Cu | Pd | 4.3–6.2 | 490–950 | $1.6 \times 10^{-6}$ | 21.9 |
| Cu | Pt | 2.4–3.5 | 490–960 | $1.0 \times 10^{-6}$ | 21.9 |
| Cu | Sn | 3.9–5.6 | 400–850 | $4.1 \times 10^{-3}$ | 31.2 |
| Cu | Zn | 6.8–9.7 | 360–880 | $3.0 \times 10^{-6}$ | 19.7 |
| Cu | Zn | 3.0 | 720–860 | $3.7 \times 10^{-6}$ | 22.0 |
| Cu | Zn | 0–9.25 | 641–884 | $5.8 \times 10^{-4}$ | 42.0 |
| Cu | Zn | 0–28.6 | 641–884 | $3.2 \times 10^{-3}$ | 42.0 |
| Cu | Zn | 27.5–35.4 | 700–950 | | 24.5 |
| Cu | Zn | α-brass | 727–955 | | 18.5 |
| Cu | Zn | β-brass | 600–720 | | 39.0 |
| Cu | Zn | 29.0 | 400–600 | | 46.0 |

*Table A.23* (continued)

| Solvent Metal | Diffusing Metal | Concentration[a] (Atom per Cent) | Temperature Interval °C | $D_0$ (cm²/sec) | Q (kcal/g atom) |
|---|---|---|---|---|---|
| Fe | α-Fe | (Self-diff.) | 715–887 | $3.4 \times 10^4$ | 77.2 |
| Fe | γ-Fe | (Self-diff.) | 935–1112 | $1.04 \times 10^{-3}$ | 48.0 |
| Fe | C | Carburization | 800–1100 | $1.67 \times 10^{-2}$ | 28.7 |
| Fe | C | 1.1 weight per cent | 900–1250 | $4.86 \times 10^{-1}$ | 36.6 |
| Fe | C | 0.1–1.0 weight per cent | 750–1250 | $0.12 \pm 0.07$ | $32.0 \pm 1.0$ |
| Pb | Pb | (Self-diff.) | 106–324 | 6.6 | 27.9 |
| Pb | Ag | <0.12 | 220–285 | $7.4 \times 10^{-2}$ | 15.2 |
| Pb | Au | Pure Au or 0.03–0.09 | 100–300 | 0.35 | 14.0 |
| Pb | Bi | 2.0 | 220–285 | $1.83 \times 10^{-2}$ | 18.4 |
| Pb | Cd | 1.0 | 167–252 | $1.83 \times 10^{-3}$ | 15.4 |
| Pb | β-Sn | 2.0 | 245–285 | 4.0 | 26.2 |
| Pb | Tl | 2.0 | 220–285 | $2.5 \times 10^{-2}$ | 19.4 |
| Pb | Tl | Various concentrations of Tl | 270–315 | 1.03 | 24.6 |
| Pt | Cu | 13.9 | 1041–1401 | $4.8 \times 10^{-2}$ | 55.7 |
| Pt | Ni | 14.9 | 1043–1401 | $7.8 \times 10^{-4}$ | 43.1 |
| W | Fe | 0.04 p.c. Fe | 1927–2527 | 11.5 | 140.0 |
| W | C | | 1702–1727 | 0.31 | 59.0 |
| W | Ce | | 1727 | 1.15 | 83.0 |
| W | Cs | 1st layer of adsorption | 27–427 | 0.2 | 14.0 |
| W | | 2nd layer of adsorption | 27–427 | 0.0164 | 2.3 |
| W | K | | 207–507 | | 15.2 |
| W | Mo | Single crystal | 1533–2260 | $6.3 \times 10^{-4}$ | 80.5 |
| W | Mo | Polycrystal | 1533–2260 | $5 \times 10^{-3}$ | 80.5 |
| W | Na | | 20–527 | 0.1 | 5.56 |
| W | Th | Grain-boundary diff. | 1780–2227 | 1.13 | 94.0 |
| W | Th | Grain-boundary diff. | 1780–2227 | 0.47 | 90.0 |
| W | Th | Volume diff. | 2127 | 1.0 | 120.0 |
| W | Th | Surface diff. | 1380 | 0.47 | 66.4 |
| W | U | | 1727 | 1.14 | 100.0 |
| W | Y | | 1727 | 0.11 | 62.0 |
| W | Zr | | 1727 | 1.1 | 78.0 |
| Zn | Zn | ‖ c Self-diff. | 355–410 | $4.6 \times 10^{-2}$ | 20.4 |
| Zn | Zn | ⊥ c Self-diff. | 340–410 | 92.0 | 31.0 |

From W. Jost, *Diffusion in Solids, Liquids, Gases,* Academic, New York, 1952.

*Table A.24   Relative Densities of Oxides, $d_{metal} = 1$*

| Metal | Li | Na | K | Rb | Cs | Mg | Ca | Ba |
|---|---|---|---|---|---|---|---|---|
| $d_{oxide}/d_{metal}$ | 0.57 | 0.58 | 0.65 | 0.46 | 0.86 | 0.85 | 0.69 | 0.71 |
| Metal | Cu | Zn | Cd | Al | Tl | Ce | Sn | Pb |
| $d_{oxide}/d_{metal}$ | 1.71 | 1.44 | 1.19 | 1.38 | 1.11 | 1.24 | 1.31 | 1.40 |
| Metal | Sb | Cr | W | Mn | Fe | Ni | Co | Pd |
| $d_{oxide}/d_{metal}$ | 1.50 | 1.97 | 3.50 | 1.75 | 2.23 | 1.64 | 1.78 | 1.60 |

From W. Jost, *Diffusion in Solids, Liquids, Gases,* Academic, New York, 1952.

Table A.25  Rate of Oxidation of Several Metals[a]

| Metal Temperature | Oxidation Constant ($g^2 \times cm^{-4} \times h^{-1}$) | | | | | | |
|---|---|---|---|---|---|---|---|
| | 408°C | 500°C | 600°C | 700°C | 800°C | 900°C | 1000°C |
| Cu in $O_2$ (173) . . . . . . | $1.64 \times 10^{-8}$ | $1.93 \times 10^{-7}$ | $1.13 \times 10^{-6}$ | $5.86 \times 10^{-5}$ | $3.14 \times 10^{-5}$ | $1.27 \times 10^{-4}$ | $6.02 \times 10^{-4}$ |
| Ni (electrolyte) in $O_2$ (173) . | | | | | $0.093 \times 10^{-6}$ | $0.76 \times 10^{-6}$ | $3.4 \times 10^{-6}$ |
| Ni ("Grade A") in $O_2$ (173) | | | | | | $1.9 \times 10^{-6}$ | $6.8 \times 10^{-6}$ |
| Fe (electrolyte) in $O_2$ (173) . | | | | $0.17 \times 10^{-4}$ | $1.00 \times 10^{-4}$ | $6.5 \times 10^{-4}$ | |
| Fe (Armco) in $O_2$ (173) . . | | | | | $1.95 \times 10^{-4}$ | $10.1 \times 10^{-4}$ | $43.0 \times 10^{-4}$ |
| Fe (Armco) in air (173) . . | | | | | $1.06 \times 10^{-4}$ | $4.9 \times 10^{-4}$ | $20.5 \times 10^{-4}$ |
| Co in air (70) . . . . . . . | | | | $5.8 \times 10^{-7}$ | $3.3 \times 10^{-6}$ | $2.2 \times 10^{-5}$ | $7.4 \times 10^{-5}$ |
| W in air (78) . . . . . . . | | | | $1.61 \times 10^{-5}$ | $\sim2 \times 10^{-4}$ | $\sim1.5 \times 10^{-4}$ | $4.61 \times 10^{-3}$ |
| Zn in $O_2$ (173) . . . . . . | $0.88 \times 10^{-10}$ | | | | | | |
| Al in $O_2$ (173) . . . . . . | | | $0.30 \times 10^{-10}$ | | | | |

[a] From C. Wagner, Handbuck du Metallphysik, I, 2, Leipzig, 1940.

# The Relation of Crystal Structure to Electronic Structure

The heats of atomization of the elements have been tabulated by Brewer,* as shown in Table B.1. They provide, as was indicated in Volume I, a measure of the bond energy between the atoms of a solid. In viewing this table it is well to remember the following points from Volume I.

1. The rare gases as solids have a low heat of atomization since they are held together by weak Van der Waals forces.

2. The strength of the bond in the halogens depends on a contribution of one electron from each atom to bind two atoms together. The molecules of the halogens form a molecular crystal when condensed to a solid.

3. Sixth group elements form two electron pair bonds.

4. Two pairs of electrons are used to bond a single pair of oxygen atoms together into a molecule.

5. The fifth group elements have three electrons missing from a completed octet (i.e., they have three single electron pair bonds).

6. Nitrogen forms a multiple bond with three pairs of electrons holding each pair of atoms together.

The foregoing list of statements indicates that bonding is stronger the greater the number of electron pairs involved in the bond. Bonding energy is roughly proportional to the number of electrons per atom or the number of electrons that are available for the formation of electron pair bonds. The average number of electrons per atom is the primary parameter of the Engel model of bonding and crystal structure in solids.

* See references 1 and 2 on p. 205.

## Table B.1    Heat of Atomization of the Solid Elements in Kilocalories/Gram Atom at 298.15° K or at the Melting Point, Whichever Temperature Is Lower[a]

| | | | | | | | | | | | | | | | | | |
|---|---|---|---|---|---|---|---|---|---|---|---|---|---|---|---|---|---|
| Li 38.4 | Be 77.9 | B 135 | | | | | | | | | | | C 170.9 | N 113.7 | O 60.4 | F 19.7 | Ne 0.50 |
| Na 25.9 | Mg 35.6 | Al 77.5 | | | | | | | | | | | Si 108 | P 79.8 | S 66 | Cl 32.2 | Ar 1.84 |
| K 21.5 | Ca 42.2 | Sc 88 | Ti 112.7 | V 123 | Cr 95 | Mn 66.7 | Fe 99.5 | Co 101.6 | Ni 102.8 | Cu 81.1 | Zn 31.2 | Ga 69 | Ge 90 | As 69 | Se 49.4 | Br 28.1 | Kr 2.55 |
| Rb 19.5 | Sr 39.1 | Y 98 | Zr 146 | Nb 173 | Mo 157.5 | Tc | Ru 155 | Rh 133 | Pd 91 | Ag 68.4 | Cd 26.75 | In 58 | Sn 72.0 | Sb 62 | Te 46 | I 25.5 | Xe 3.57 |
| Cs 18.7 | Ba 42.5 | La 102 | Hf 160 | Ta 186.8 | W 200 | Re 187 | Os 187 | Ir 155 | Pt 135.2 | Au 87.3 | Hg 15.32 | Tl 43.0 | Pb 46.8 | Bi 49.5 | Po 34.5 | At | Rn |
| Fr | Ra | Ac | | | | | | | | | | | | | | | |

| Pr | Nd | Pm | Sm | Eu | Gd | Tb | Dy | Ho | Er | Tm | Yb | Lu |
|---|---|---|---|---|---|---|---|---|---|---|---|---|
| 80 | 77 | | 50 | 42 | 84 | 80 | 62 | 70 | 66 | 58 | 40 | 95 |
| ±4 | ±1 | | ±3 | ±3 | ±2 | ±7 | ±4 | ±3 | ±4 | ±2 | ±5 | ±10 |
| Pa | U | Np | Pu | Am | Cm | Bk | Cf | Es | Fm | Md | | Lw |
| 126 | 125 | 105 | 92 | 66 | | | | | | | | |

[a] From L. Brewer (see references 1 and 2, p. 205 of this text).

## Table B.2   Crystal Structures of the Elements[a]

| 1 | 2 | 3 | 4 | 5 | 6 | 7 | 8 | 9 | 10 | 11 | 12 | 13 | 14 | 15 | 16 | 17 | 18 |
|---|---|---|---|---|---|---|---|---|---|---|---|---|---|---|---|---|---|
| Li<br>I | Be<br>*, II | B<br>*, * |  |  |  |  |  |  |  |  |  |  | C<br>IV, $\geq$f | N<br>$\equiv$b | O<br>$=$b | F<br>—c | Ne |
| Na<br>I | Mg<br>II | Al<br>III |  |  |  |  |  |  |  |  |  |  | Si<br>IV | P<br>↘e | S<br>&lt;d | Cl<br>—c | Ar |
| K<br>I | Ca<br>I, III | Sc<br>II | Ti<br>I, II | V<br>I | Cr<br>III, I | Mn<br>I, III, ** | Fe<br>I, III, I | Co<br>III, II | Ni<br>III | Cu<br>III | Zn<br>II | Ga<br>* | Ge<br>IV | As<br>↘e | Se<br>&lt;d | Br<br>—c | Kr |
| Rb<br>I | Sr<br>I, II, III | Y<br>II | Zr<br>I, II | Nb<br>I | Mo<br>I | Tc<br>II | Ru<br>II | Rh<br>III | Pd<br>III | Ag<br>III | Cd<br>II | In<br>(III)g | Sn<br>*, IV | Sb<br>↘e | Te<br>&lt;d | I<br>—c | Xe |
| Cs<br>I | Ba<br>I | La<br>III, II | Hf<br>I, II | Ta<br>I | W<br>I | Re<br>II | Os<br>II | Ir<br>III | Pt<br>III | Au<br>III | Hg<br>* | Tl<br>I, II | Pb<br>III | Bi<br>↘e | Po | At | Rn |
| Fr<br>I | Ra<br>I | Ac<br>II |  |  |  |  |  |  |  |  |  |  |  |  |  |  |  |

| Ce | Pr | Nd | Pm | Sm | Eu | Gd | Tb | Dy | Ho | Er | Tm | Yb | Lu |
|---|---|---|---|---|---|---|---|---|---|---|---|---|---|
| ?<br>III | III?<br>II | ?<br>II |  | (II)g | I | II | II | II | II | II | II | III | II |
| Th<br>I, III | Pa<br>* | U<br>I, ** | Np<br>I?, ** | Pu<br>I, *, III, ** | Am | Cm | Bk | Cf | Es | Fm | Md |  | Lw |

[a] I: Body centered cubic. II: Hexagonal close packed. III: Cubic close packed. IV: Diamond. Asterisk denotes complex.

[b] Diatomic molecules with double or triple bonds.

[c] Diatomic molecule with a single electron pair bond.

[d] Atoms which form two single bonds per atom to form rings or infinite chains.

[e] Three single bonds per atom, corresponding to a puckered planar structure.

[f] The graphite structure where one resonance form consists of two single and one double bond per atom.

[g] Parentheses indicate slight distortions.

From L. Brewer (see references 1 and 2, p. 205 of this text).

*Table B.3   The Engel Correlation Between Crystal Structure and Electron Configuration[a]*

---

1. The stability or bonding energy of a solid depends upon the average number of unpaired electrons per atom available for bonding.
2. Contribution of the *d* electrons to bonding increases with increasing atomic number of the metals in the periodic table, whereas the contribution of the *sp* electrons decreases.
3. The crystal structure of a solid depends on the number of *s* and *p* electrons present.
    (a) BCC for <1.5 *sp* electrons per atom.
    (b) HCP for 1.7 − 2.1 *sp* electrons per atom.
    (c) FCC for 2.5 − 3+ *sp* electrons per atom.
    (d) *d* electrons do not directly fix the crystal structure.
4. The number of *s, p, d, f* electrons in metal phases is close to that for gaseous atoms. If neighboring atoms have unpaired *d* electrons available for bonding, their number of unpaired *d* electrons is higher.
5. Carbon and nitrogen in combination with transition metals of the IV and V groups of the periodic table convert the electronic structure to that of the more stable VI group elements.

---

[a] From L. Brewer (see references 1 and 2, p. 205 of this text).

To independently confirm the Engel bonding model for predicting the crystal structure of alloy phases, Brewer has provided an independent confirmation. He has taken the spectroscopic data pertaining to the electron configuration for each state. In potassium, for example, the ²S state indicates an electron configuration outside the closed shell of a single *s* electron. The ²P state is the first excited state; its excitation energy is 37 kcals/mole and involves a single *p* electron. The ²D state is 62 kcals/mole of energy above the ground state ²S. Thus to raise (promote) the *s* electron of the ground state of potassium to a *d* electron requires 62 kcals/mole of energy. Such configurations provide only one electron per atom for bonding.

In Engel's scheme two electronic configurations with the same number of unpaired electrons used for bonding purposes—such as $d^5s$ and $d^4sp$, for example—give equivalent bonding energies. If one of these requires a higher promotional energy than the other, only the one of lower energy should be considered as stable.

*Table B.4   Promotion Energies and Bonding Energy for Metals in K-calories/mole\**

| | | $d^5s(^7S)$ | $d^4s^2(^5D)$ | $d^4sp(^7F)$ | $d^5s$ bond | $\Delta H_{subl}$ |
|---|---|---|---|---|---|---|
| Group VI | Cr | 0 | 23 | 70 | 95 | 95 |
| | Mo | 0 | 31 | 80 | 158 | 158 |
| | W | 9 | 0 | 55 | 209 | 200 |

| | | $d^3s^2(^4F)$ | $d^4s(^6D)$ | $d^3sp(^6G)$ | $d^4s$ bond | $\Delta H_{subl}$ |
|---|---|---|---|---|---|---|
| Group V | V | 0 | 6 | 47 | 129 | 123 |
| | Nb | 0 | 3 | 47 | 176 | 173 |
| | Ta | 0 | 28 | 50 | 215 | 187 |

| | | $d^2s^2(^3F)$ | $d^3s(^5F)$ | $d^2sp(^5G)$ | $d^3s$ bond | $d^2sp$ bond |
|---|---|---|---|---|---|---|
| Group IV | Ti | 0 | 19 | 45 | 132 | 158 |
| | Zr | 0 | 14 | 42 | 160 | 188 |
| | Hf | 0 | 40 | 51 | 200 | 211 |

| | | $s^2p(^2P)$ | $ds^2(^2D)$ | $d^2s(^4F)$ | $dsp(^4F)$ | $sp^2(^4P)$ |
|---|---|---|---|---|---|---|
| Group III | Al | 0 | 92 | | 190 | 83 |
| | Sc | 97 | 0 | 33 | 45 | |
| | Y | 30 | 0 | 31 | 43 | |
| | La | 44 | 0 | 6 | 40 | |

| | | $s^2$ | $sp$ | $sd$ |
|---|---|---|---|---|
| Alkaline earths | Mg | 0 | 80 | 140 |
| | Ca | 0 | 55 | 60 |
| | Sr | 0 | 52 | 56 |
| | Ba | 0 | 44 | 29 |

| | | $d^5s^2(^6S)$ | $d^6s(^6D)$ | $d^5sp(^8P)$ | $d^5sp$ bond |
|---|---|---|---|---|---|
| Group VII | Mn | 0 | 49 | 53 | 120 |
| | Tc | 0 | 6 | 46 | 202 |
| | Re | 0 | 35 | 56 | 243 |

| | | $d^6s^2(^5D)$ | $d^7s(^5F)$ | $d^6sp(^7D)$ | $d^6sp$ bond |
|---|---|---|---|---|---|
| Group VIII | Fe | 0 | 22 | 57 | <157 |
| | Ru | 19 | 0 | 70 | 223 |
| | Os | 0 | 15 | 70 | 257 |

\* See Table B.1. From L. Brewer (see references 1 and 2, p. 205 of this text).

In Table B.4 the promotion energies of Cr, Mo, and W head the list. The first three columns list the energies of the lowest electronic states ($^7S$, $^5D$, $^7F$), which have the electronic configurations $d^5s$, $d^4s^2$, and $d^4sp$ respectively. The promotion energies indicate that a BCC structure arising from the $d^5s$ configuration is more stable than an HCP structure arising from the $d^4sp$ configuration. The $d^4s^2$ does not count, because the paired $s$ electrons in this case are not available, and the configuration yields only four electrons for bonding each atom. The other two configurations can both furnish six electrons. FCC cubic structure is ruled out because of the very high promotion energy required.

The fifth column, listed only for group IV and V elements, tabulates the enthalpy in kcal/mole which is liberated when gaseous atoms in the $^7S(d^5s)$ state condense to form a solid. These values have been obtained by combining enthalpies of sublimation at 298°K with promotion energies at 0°K. An HCP structure might be expected for Cr, Mo, and W if the six bonding electrons ($d^4sp$ configuration) had energies of 165, 238, and 255 kcal/mole. If we use the six electrons in the $d^5s$ configuration, the bonding energies would be 95, 158, and 209 kcal/mole. Thus the spectroscopic data for free gaseous atoms is sufficient, according to Engel, to indicate that the observed BCC structure is preferred.

For group V elements, the $d^4s$ configuration with five bonding electrons is preferred for V and Nb. For group V and VI transition metals, Engel's rule indicates $d^4s$ and $d^5s$ configurations respectively and BCC structures. For group IV metals with the configuration $d^3s$ and $d^2sp$ with four unpaired electrons, BCC and HCP structures are possible.

Among the group III metals, Al is an interesting example. Although no data are available for the $d^2s$ configuration, it is expected to be high; hence a BCC structure is ruled out. An HCP structure corresponding to a configuration $dsp$ is too high by 100 kcal/mole; thus $sp^2$ corresponding to a FCC structure is the only one possible. This is in accord with the experimental x-ray findings, as was so carefully pointed out by Brewer and which is summarized in Table B.4.

Mixing of metals in the eighteen metals of the first six groups of the three transition series does not change the number of $sp$ elec-

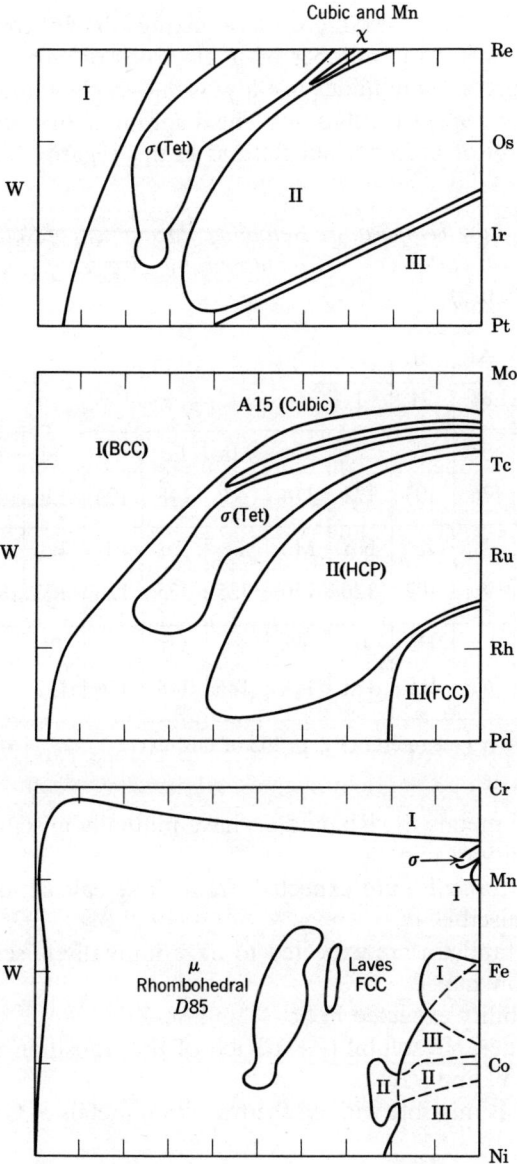

Figure B.1    Phase regions which follow Engel correlation, according to Brewer. Experimental verification exists for most of the regions shown. From L. Brewer (see reference 2, p. 205 of this text).

trons and may be viewed merely as a mixing of *d* electrons. Thus, lack of change in the number of *sp* electrons on mixing permits Brewer to predict solubilities in alloys using simple solution theory to check the Engel Correlation. The diagrams shown in Figure 1 are the result of Brewer's calculations in this regard.

*Table B.5    Low Temperature Solubility Parameters of Solid Metals,* $(\Delta E/V)^{1/2}$ *in* $cal^{1/2}/cc^{1/2}$ *(Calculated by Brewer from Regular Solution Theory)*

| Na | Mg | Al | Si | | | | | | | | |
|------|------|------|------|------|------|------|------|------|------|------|------|
| 33 | 50 | 88 | 94 | | | | | | | | |
| K | Ca | Sc | Ti | V | Cr | Mn | Fe | Co | Ni | Cu | Zn |
| 21.5 | 40 | 76 | 103 | 121 | 114 | 95 | 118 | 124 | 124 | 107 | 58 |
| Rb | Sr | Y | Zr | Nb | Mo | Tc | Ru | Rh | Pd | Ag | Cd |
| 18.5 | 34 | 70 | 102 | 126 | 130 | 135 | 135 | 126 | 101 | 81.5 | 45 |
| Cs | Ba | La | Hf | Ta | W | Re | Os | Ir | Pt | Au | Hg |
| 16 | 34 | 67 | 108 | 131 | 146 | 146 | 148 | 136 | 122 | 92 | 32 |

From L. Brewer (see reference 2, p. 205 of this text).

1. Alkali metals, K, Rb, and Cs have mutually miscible phases in the vicinity of m.p.
2. Ca, Sr, and Ba are expected, from these calculations, to be mutually miscible.
3. Si, Y, and Ca are expected to have mutually miscible BCC phases below m.p.
4. Miscibility expected in Sc–Ti and Sc–Zr.
5. Considerable solubility expected of the transition metals of Group IV, V, and VI.

Solid Nb is miscible with all the transition metals of Groups IV, V, and VI.

Solid Mo is miscible with all except Zr and Hf.

Solid W is miscible only with Mo, Cr, Ta, and Nb.

Solid Ta behaves like W except that it shows higher solubility.
Solid Ti is abnormally soluble.
Solid Zr shows less solubility.

6. Brewer has also discussed solid phases other than BCC solid solutions.

REFERENCES

1. Brewer, L., *Electronic Structure and Alloy Chemistry of the Transition Elements,* Ed. P. A. Beck, Wiley, New York, 1962. "Thermodynamic Stability and Bond Character in Relation to Electronic Structure and Crystal Structure" introduces Engel model.
2. Brewer, L., *Prediction of High Temperature Metallic Phase Diagrams,* University of California Radiation Laboratory Report, Berkeley, **10701,** July, 1963. Uses regular solution theory to justify Engel model.
3. Engel, N., *Kem Maanedsblad,* **5, 6, 8, 9,** and **10** (1949). Danish.
4. Engel, N., *Planseeberichte and Powder Metallurgy Bulletin,* **7, 8** (1954). Summarizes the original. (See preceding reference.)

# *Ternary Diagrams*

Most industrial alloys contain one major constituent, another in moderate concentration, and several more as purposeful or accidental additions.  A binary alloy diagram is therefore hardly sufficient in describing any ternary phases present.  Consequently, an elementary understanding of the principles governing ternary phases is useful.  Multicomponent phases of greater complexity can hardly be handled diagramatically.

By fixing the pressure constant, we can use a triangular prism phase diagram of a ternary system.  Figure C.1 indicates the

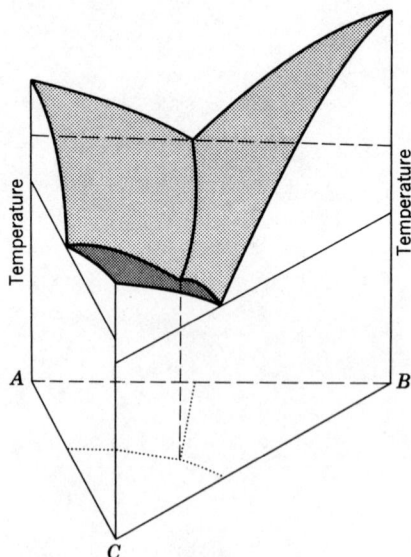

Figure C.1   Ternary diagram of three binary eutectic systems, none showing intermediate planes.  Adapted from C. E. Birchenall, *Physical Metallurgy*, McGraw-Hill, New York, 1959.

206

liquidus of an oversimplified ternary phase diagram made up of three eutectics without any solubility. Isothermal sections of this ternary diagram can be broken up, as in Figure C.2a, for some temperature below the melting point of metal B. Figure C.2b represents an isothermal section below the binary AB eutectic temperature. Figure C.2c depicts an isothermal section just above the ternary eutectic temperature. The solid lines in Figures C.2a, b, and c represent the intersection of isothermal planes with the liquidus surfaces. The dotted lines indicate the boundaries of the three phase fields.

For some purposes it can be more useful to project the liquidus surface or solidus surface on the base triangle. This is shown in Figure C.1 for the liquidus surfaces intersecting as dotted lines on the base triangle. Another useful plot of a ternary diagram is the *pseudobinary* section in which the concentration of one component is held constant and the proportions of the other two are varied as in a binary system.

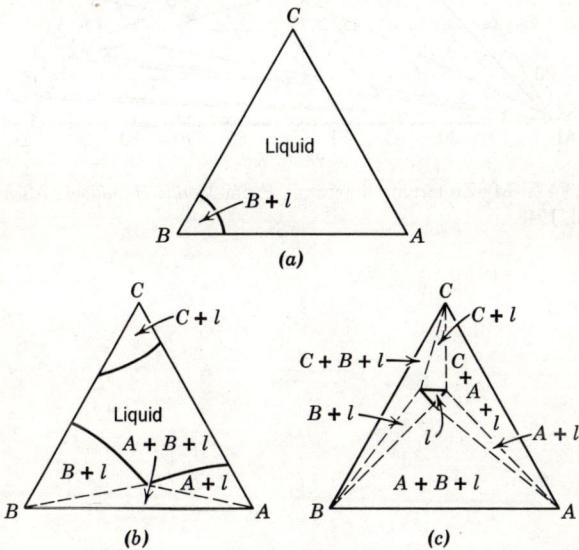

Figure C.2   Isothermal sections of Figure C.1.   (a) Just below melting point of B. (b) Just below AB eutectic temperature.   (c) Just above ternary eutectic temperature. Adapted from C. E. Birchenall, *Physical Metallurgy*, McGraw-Hill, New York, 1959.

Figure C.3 depicts an isothermal section at 100°C of the Al-Mg-Zn system. Figure C.4 shows an isothermal section of the Fe-Ni-Cr phase diagram. The phases with the same structure are designated by the same letter. In this case the α phases are BCC, the γ phase is FCC, and the σ has a complex tetragonal structure.

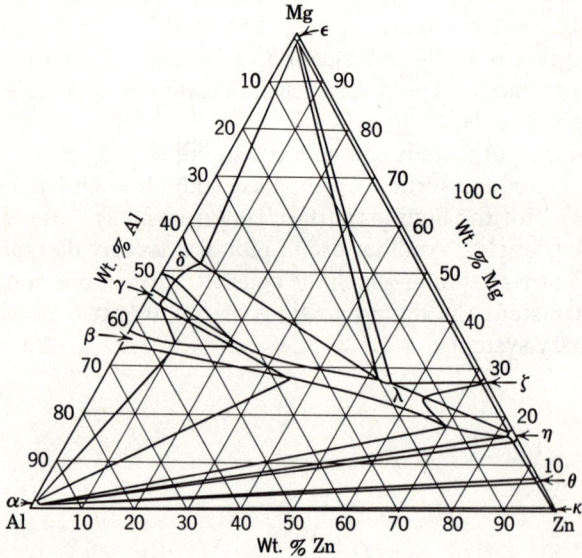

Figure C.3    Al-Mg-Zn ternary diagram.    From *Metals Handbook,* ASM, p. 1247, Cleveland, 1948.

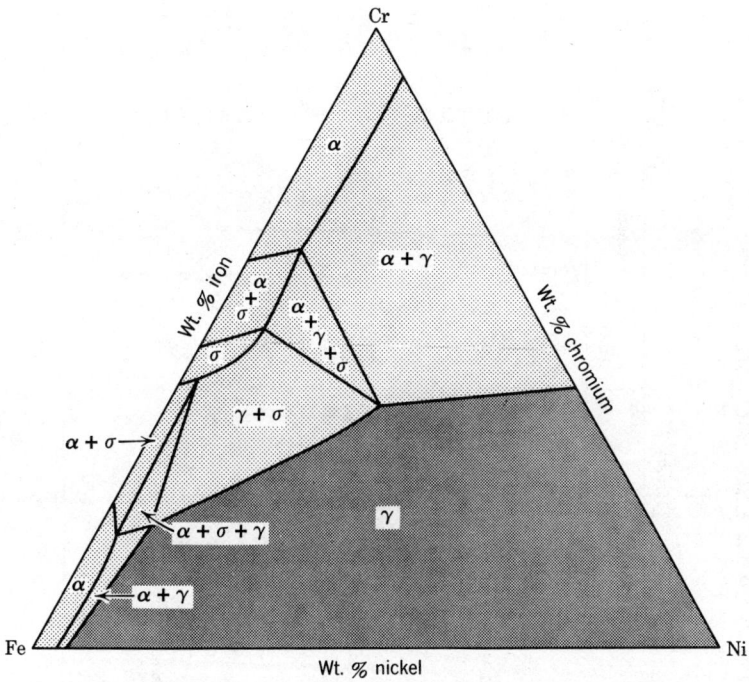

Figure C.4   Fe-Ni-Cr ternary phase diagram.   From the *Metals Handbook,* ASM, p. 1261, Cleveland, 1948.

## Table C.1 Melting Temperatures and Invariant Points for the CaO-Al$_2$O$_3$-SiO$_2$ System[c]

| Compound | Melting Temperature °C | Phase Name | Composition Al$_2$O$_3$ | CaO | SiO$_2$ |
|---|---|---|---|---|---|
| A-Al$_2$O$_3$ | ~2020 | Corundum | 100 | — | — |
| C-CaO | ~2570 | Lime | — | 100 | — |
| S-SiO$_2$ | 1723 | Cristobalite | — | — | 100 |
| S-SiO$_2$ | | Tridymite | — | — | 100 |
| S-SiO$_2$ | | Quartz[a] | — | — | 100 |
| CS | 1544 | Wollastonite[a] | — | 48 | 52 |
| A$_3$S$_2$ | ~1850 | Mullite | 72 | — | 28 |
| CAS$_2$ | 1553 | Anorthite | 37 | 20 | 43 |
| C$_2$AS | 1593 | Gehlenite | 37 | 41 | 22 |
| C$_3$S$_2$ | D[b] | Rankinite | — | 58 | 42 |
| C$_2$S | ~2130 | Lamite[a] | — | 65 | 35 |
| C$_3$S | D[b] | — | — | 74 | 26 |
| C$_3$A | D[b] | — | 38 | 62 | — |
| C$_{12}$A$_7$ | 1455 | — | 51 | 49 | — |
| CA | ~1605 | — | 64.5 | 35.5 | — |
| CA$_2$ | ~1750 | — | 78.5 | 21.5 | — |
| CA$_6$ | D[b] | — | 91.5 | 8.5 | — |

### Invariant Points

**Ternary**

| | Temperature °C | Al$_2$O$_3$ | CaO | SiO$_2$ |
|---|---|---|---|---|
| 1 | 1470 | 12 | 5 | 83 |
| 2 | 1345 | 20 | 10 | 70 |
| 3 | 1512 | 37 | 15 | 48 |
| 4 | 1495 | 41 | 23 | 36 |
| 5 | 1380 | 39 | 29 | 32 |
| 6 | 1475 | 45 | 31 | 24 |
| 7 | 1500 | 53 | 37.5 | 9.5 |
| 8 | 1380 | 42 | 48.5 | 9.5 |
| 9 | 1335 | 43.5 | 49.5 | 7 |
| 10 | 1335 | 41.5 | 52 | 6.5 |
| 11 | 1455 | 33 | 58.5 | 8.5 |
| 12 | 1470 | 33 | 59.5 | 7.5 |
| 13 | ~1315 | 12 | 49 | 39 |
| 14 | ~1310 | 12 | 47 | 41 |
| 15 | 1266 | 20 | 38 | 42 |
| 16 | 1170 | 15 | 23 | 62 |

**Binary**

| | Temperature °C | Al$_2$O$_3$ | CaO | SiO$_2$ |
|---|---|---|---|---|
| A | ~1590 | 6 | — | 94 |
| B | ~1840 | 78 | — | 22 |
| C | ~1850 | 89 | 11 | — |
| D | ~1730 | 80 | 20 | — |
| E | ~1595 | 66.5 | 33.5 | — |
| F | 1400 | 53 | 47 | — |
| G | 1395 | 50 | 50 | — |
| H | 1535 | 41 | 59 | — |
| I | ~2070 | — | 72 | 28 |
| J | ~2050 | — | 70 | 30 |
| K | 1464 | — | 55.5 | 44.5 |
| L | 1460 | — | 55 | 45 |
| M | 1436 | — | 36 | 64 |
| N | 1470 | — | 35 | 65 |
| O | 1707 | — | 27 | 73 |
| P | 1707 | — | 1 | 99 |
| Q | 1368 | 19.5 | 11.5 | 70 |
| R | 1547 | 39.5 | 18.5 | 42 |
| S | 1552 | 50 | 35 | 15 |
| T | 1512 | 52.5 | 38 | 9.5 |
| U | 1335 | 41 | 52.5 | 6.5 |
| V | 1545 | 24 | 50 | 26 |
| W | 1318 | 13 | 46 | 41 |
| Y | 1307 | 37 | 30 | 33 |
| Z | 1385 | 19 | 34 | 47 |

[a] Name of 25°C polymorph.
[b] D = decomposes.
[c] Shown in Figure C.5 according to adaptation of L. H. Van Vlack, *Elements of Materials*, Addison Wesley, Massachusetts, 1964.

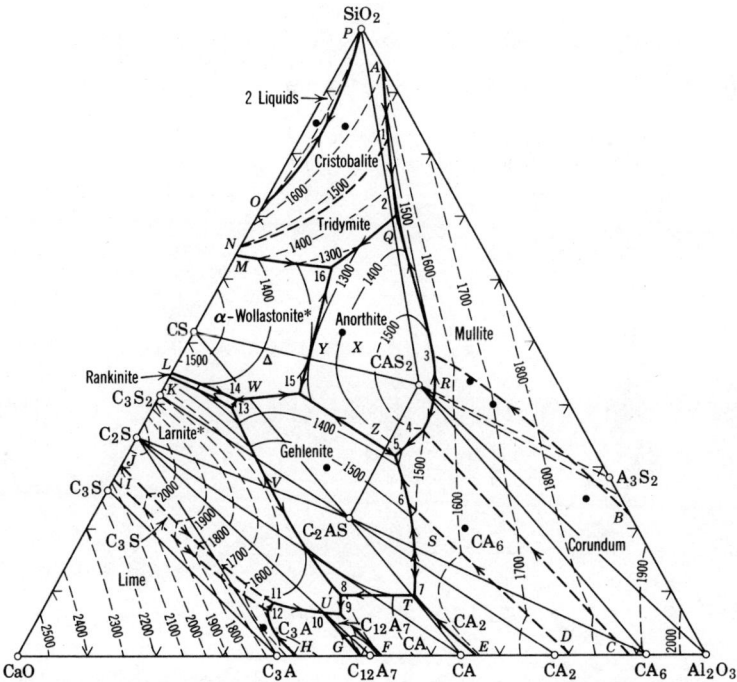

Figure C.5   The CaO-Al₂O₃-SiO₂ system.   (Adapted, with some simplifications, by
H. Van Vlack from E. F. Osborn and A. Muan, "The System CaO-Al₂O₃-SiO₂," *Phase
Equilibrium Diagrams of Oxide Systems,* Columbus, Ohio, American Ceramic Society,
Plate I, 1960.

# Index